tear here

The Complete Idiot's Guide Visual InterDev Tear-

Data Types

Table A Types of variant data types

DATA SUBTYPE	*DESCRIPTION*
Empty	Variant is uninitialized. If the context uses it as a number, the value is zero. If the context uses it as a string it is an empty or zero-length string (" ").
Null	Variant contains no valid data. Nulls can be confusing. It does not mean blank. It means we don't know and can't know the value. A null in a logical comparison will not match any value.
Boolean	The value is True or False, on or off, yes or no, one or zero. This can be represented in memory by one bit.
Byte	The value is an integer in the range of 0 to 255. This is not a signed value. This means that it is not positive of negative.
Integer	The value is a signed integer in the range of - (negative) 32,768 to 32,768 to + (positive) 32,768 to 32,767. Note that there is no unsigned integer as there is some other languages.
Long	The value is a long integer in the range of -2,147,483,648 to +2,147,483,647.
Currency	922,337,203,685,477.5808 to 922,337,203,685,477.5807 is the range of values.
Single	The values are a single-precision, floating point number in the range of -3.402823E38 to -1.401298E-45 for negative values to 1.401298E-45 to 3.402823E38 for positive values.
Double	The values are a double-precision, floating point number in the range of -1.79769313486232E308 to -4.94065645841247E-324 for negative values to 4.94065645841247E-324 to 1.79769313486232E308 for positive values.
Date (Time)	The value is a number that represents a date and time between January 1, 100 to December 31, 9999.
String	Contains a variable length string that can be roughly 2 billion characters long.
Object	Contains an object. This includes Err objects, Dictionary objects which are a keyed pair array, the FileSystemObject object which provides access to the computer's files, and the TextStream object which is used to read sequential files.
Error	Contains an error number.

Functions

Table B To determine the subtype of a variable

FUNCTION NAME	*FUNCTION DESCRIPTION*
VarType(varname)	Returns a numeric value depending on the Variant subtype, for example if the function returns the number 2 it is an integer and if it returns 8 the variable is a string.
TypeName(varname)	Returns a string that tells the variant subtype, for example "byte" or "string".

Table C To test for specific Variant subtypes

FUNCTION NAME	*FUNCTION DESCRIPTION*
IsArray(varname)	Returns a true or false value depending on whether the variable is an array.
IsNull(varname)	Returns a true or false value depending on whether the variable value is null or not.

continues

continued

Table D Functions that convert Variant subtypes, which will leave you with a known subtype:

FUNCTION NAME	*FUNCTION DESCRIPTION*
Cint(expression)	Returns a value of the integer subtype.
CStr(expression)	Returns a value in the form of a string subtype.

Table E VBScript Arithmetic Operators

OPERATOR	*NAME*	*EXAMPLE*
+	Addition	X = A + B
-	Subtraction	X = A - B
*	Multiplication	X = A * B
/	Division	X = A/B
\	Integer Division	X = A\B (X will be an integer)
^	Exponentation	X = A^B (A to the B power)
Mod	Modulus Arithmetic	X = A Mod B (X is the remainder after A is divided by B)

Table F Comparison Operators

OPERATOR	*NAME*	*EXAMPLE*
=	Equality	A = B
>	Greater than	A > B
<	Less than	A < B
<>	Not equal to	A <> B
>=	Greater than or equal to	A >= B
<=	Less than or equal to	A <= B
Is	Tests to see if two objects are the same object.	ObjectA Is ObjectB

Table G Logical Operators

OPERATOR	*NAME*	*EXAMPLE*
And	Logical conjunction	A and B
Not	Logical negation	Not A
Or	Logical disjunction	A or B (or both)
Xor	Logical exclusion	A or B (but not both)
Eqv	Logical equivalence	A is equivalent to B
Imp	Logical implication	A implies B

Table H Suggested prefixes of variables based on data subtype:

PREFIX	*MEANING*	*EXAMPLE*	*PREFIX*	*MEANING*	*EXAMPLE*
bln	Boolean	blnMyBool	int	Integer	intMyNumber
byt	Byte	bytMyValue	lng	Long	lngMyBigInteger
cur	Currency	curSomeMoney	obj	Object	objMyObject
dtm	Date/Time	dtmBirthday	sng	Single	sngMyDecimalNumber
dbl	Double	dblBigNumber	str	String	strMyFirstName
err	Error	errMyError			

Table I Script Computation Operators

OPERATOR	*NAME*	*EXAMPLE*
+	Addition	X = A + B
-	Subtraction	X = A - B
*	Multiplication	X = A * B
/	Division	X = A/B
%	Modulus Arithmetic	X = A Mod B (X is the remainder after A is divided by B)
++	Increment	5++ will be raised by 1 to 6
--	Decrement	5-- will be reduced by 1 to 4

Table J Script Logical Operators

OPERATOR	*NAME*	*EXAMPLE*
==	Equality	A == B
>	Greater than	A > B
<	Less than	A < B
!=	Not equal to	A != B
>=	Greater than or equal to	A >= B
<=	Less than or equal to	A <= B
!	Logical NOT Variable !Expression	If the expression evaluates to 0 then Variable will be True.
&&	Logical AND Variable = Expression1 && Expression2	If both expressions are true then Variable is True
||	Logical OR Variable = Expression1 || Expression2	If either expression is true then Variable is True

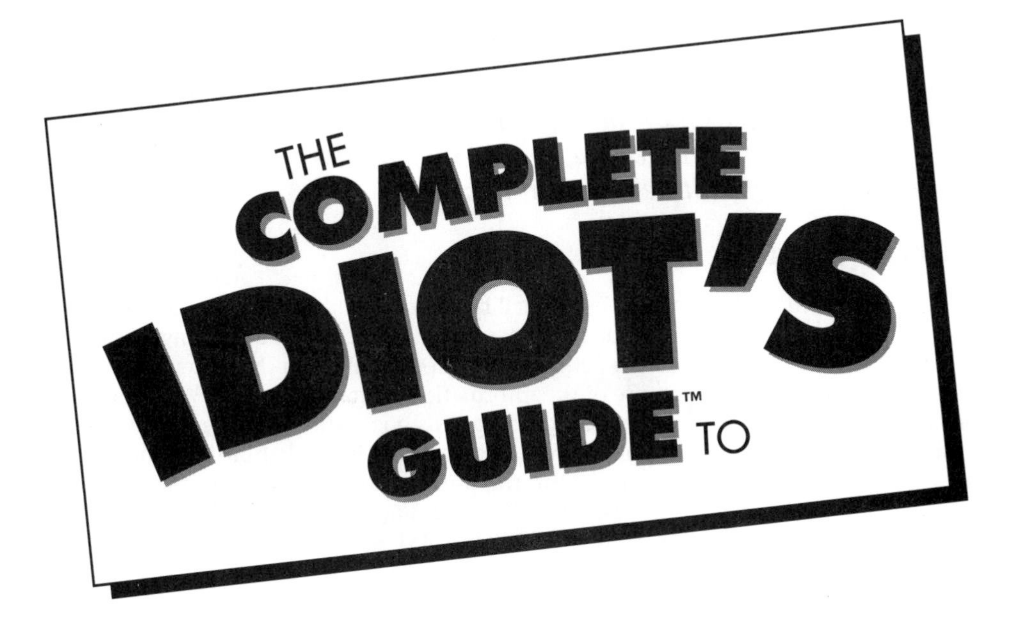

Microsoft Visual InterDev

by Nelson Howell

A Division of Macmillan Computer Publishing
201 W.103rd St., Indianapolis, IN 46290 USA

To my partner Wanda who has the patience of Job, the beauty of Venus, and my eternal devotion; to my sons Matt, Drew, Chris, and James who keep me challenged; and to my parents Virgil and Velma Howell who never quit believing in me.

 For information, address Que Corporation, 201 West 103rd Street, Indianapolis, IN 46290. You can reach Que's direct sales line by calling 1-800-428-5331.

Library of Congress Catalog Card Number: 93-56761-168-0

International Standard Book Number: 0-7897-1217-2

99 98 97 8 7 6 5 4 3 2 1

Interpretation of the printing code: the rightmost double-digit number is the year of the book's first printing; the rightmost single-digit number is the number of the book's printing. For example, a printing code of 97-1 shows that this copy of the book was printed during the first printing of the book in 1997.

Screen reproductions in this book were created by means of the program Collage Complete from Inner Media, Inc., Hollis, NH.

Printed in the United States of America

Publisher
Roland Elgey

Publishing Director
Lynn E. Zingraf

Editorial Services Director
Elizabeth Keaffaber

Managing Editor
Michael Cunningham

Director of Marketing
Lynn E. Zingraf

Acquisitions Editor
Martha O'Sullivan

Technical Specialist
Nadeem Muhammed

Product Development Specialist
John Gosney

Technical Editors
Suni Hazari, Matt Brown

Production Editor
Tom Lamoureux

Illustrator
Judd Winick

Copy Editors
Nick Zafran, Kate Givens

Book Designer
Kim Scott

Cover Designers
Dan Armstrong, Barbara Kordesh

Production Team
Tricia Flodder, Daniela Raderstorf,
Rowena Rappaport, Megan Wade

Indexer
Greg Pearson

We'd Like to Hear from You!

As part of our continuing effort to produce books of the highest possible quality, Que would like to hear your comments. To stay competitive, we *really* want you, as a computer book reader and user, to let us know what you like or dislike most about this book or other Que products.

You can mail comments, ideas, or suggestions for improving future editions to the address below, or send us a fax at (317) 581-4663. For the online inclined, Macmillan Computer Publishing has a forum on CompuServe (type **GO QUEBOOKS** at any prompt) through which our staff and authors are available for questions and comments. The address of our Internet site is **http://www.mcp.com/que** (World Wide Web).

In addition to exploring our forum, please feel free to contact me personally to discuss your opinions of this book: I'm **104436,2300** on CompuServe, and I'm **jgosney@que.mcp.com** on the Internet.

Thanks in advance—your comments will help us to continue publishing the best books available on computer topics in today's market.

John Gosney
Product Development Specialist
Que Corporation
201 W. 103rd Street
Indianapolis, Indiana 46290
USA

Although we cannot provide general technical support, we're happy to help you resolve problems you encounter related to our books, disks, or other products. If you need such assistance, please contact our Tech Support department at 800-545-5914 ext. 3833.

To order other Que or Macmillan Computer Publishing books or products, please call our Customer Service department at 800-835-3202 ext. 666.

Acknowledgments

I would like to thank Steve Wynkoop, a good friend and the person most to blame for my attempting this book. Your guilt looms large.

Thanks go to Martha O'Sullivan, John Gosney, and Tom Lamoureux at Que for their patience and understanding. They were always there when I needed help.

Special thanks to Angela Kozlowski, Kelly Marshall, Al Valvano, and Carmen Krikorian of Que who helped me with my first toddling steps as a writer.

I would like to thank the hundreds of people with whom I have worked in the computer industry over the last 30 years. You have kept my life from being boring.

And a large helping of gratitude to my wife Wanda and my son James who tolerated the long hours of listening to my muttering and typing.

Trademarks

Contents at a Glance

Part 1: What Is Microsoft Visual InterDev, and How Does It Open Internet Technology? **1**

1 Internet Technologies in Perspective: Some History and Facts 3
Where we have been helps us understand where we are going.

2 Internet Versus an Intranet—More Than Just a Typo 9
Internet technology works both for public and private applications.

3 World Wide Web + Emerging Standards = Chaos 17
You will want to understand what the various standards are and who creates and amends them.

4 Microsoft Visual InterDev: Order From Chaos 23
Visual InterDev helps sort out the confusion and provides tools to master the environment.

Part 2: Understanding the Visual InterDev Interface **29**

5 Toolbars/Menus/Help and Documentation: What Does This Do? 31
Navigating Visual InterDev and setting it up to suit your needs is simple.

6 Creating and Editing Workspaces, Projects, and Files 45
Understanding the Workspace and Projects will help you make effective use of Visual InterDev.

Part 3: Active Server Pages **63**

7 Active Server Pages—Don't Just Sit There! 65
Active Server brings a new dimension to Web sites, allowing indepence from browser capabilities.

8 VBScript: Lights, Camera, Action! 77
VBScript is an easy-to-use language that brings life to your Active Server Pages.

9 Java, JavaScript, Java Applets: More Than Just a Cup of Coffee 91
JavaScript is an alternative to VBScript but is not directly related to Java.

10 ActiveX Controls, OLE & VB5CCE: Making Your Own 105
Microsoft put OLE on a diet and created ActiveX Controls. You can also create your own easily with Visual Basic 5.0.

11 Multi-Tier Applications 115
Separating and controlling User Interface, Business Rules, and Databases has never been easier.

Part 4: Databases **121**

12 Relational Databases and SQL: Information, Please! 123
A common language and syntax, SQL, is used to communicate with relational databases.

13 Query Designer, or Instant SQL 135
Visual InterDev provides a tool to create correct SQL statements with ease.

14 Database Wizards, or Instant Databases 149
Visual InterDev provides a tool to build SQL Server 6.5 databases more easily than ever.

15 Database Connectivity—Long Distance, Please 159
The set of middleware called ODBC (Open Database Connectivity) provides the connection between the user and the database.

Part 5: Integrating Microsoft Visual InterDev **171**

16 Working with Link View: Where Is That Page? 173
Link view presents a graphic view of a Web site and tests the links.

17 Working with FrontPage 97: What You See 181
You have the FrontPage Editor available to perform WYSIWYG editing of HTML pages. Projects can be jointly developed by users of Visual InterDev and FrontPage.

18 Source Code Security/Visual SourceSafe: You Changed What? 189
Visual InterDev integrates with Visual SourceSafe 5.0 to provide complete security for you source code.

19 Web Site Security: Guarding the Gate 203
How do you keep visitors to you Web site within the bounds that you have set? There are several mechanisms that work together.

20 Client Scripting—No, You Do It 209
It can be very useful to move functionality to the client for editing and validation of data during data entry.

21 HTML Layout Control—Put It Where? 219
The ActiveX HTML layout control provides precise control of the location of objects on the client page.

22 Internet Explorer Script Debugger: It did What? 229
Microsoft has provided a very useful script debugger that runs under the Internet Explorer.

Part 6: Finishing Touches **235**

23 Arts and Crafts: Media Manager, Image Composer, and Music Producer 237
Microsoft provides useful tools for creating royalty free images and music multimedia files for your Web pages.

24 Creating Your First Sample Project: Where Are the Blueprints? 247
The best way to learn is to do. Create a functional Web Site as practice.

Part 7: Final Issues **265**

25 My Web Server's Better Than Your Web Server 267
Whether you use MS IIS or Personal Web Server, you will want to know the difference.

Appendix A: Installation and Setup of Software 275
How to install all of the software.

Appendix B: What's on the CD? 281
A review of the contents of the companion CD.

Appendix C: Speak Like a Geek: The Complete Archive 283
A translation of all of the terms known only to "Computer Geeks."

Index 293

Contents

Part 1: What Is Microsoft Visual InterDev, and How Does It Open Internet Web Technology? 1

1 Internet Technologies in Perspective: Some History and Facts 3

Internet Services ... 4
Transfer Connect Protocol/Internet Protocol (TCP/IP) ... 6
TCP (Transfer Connect Protocol) ... 6
IP (Internet Protocol) ... 7
The Least You Need to Know ... 7

2 Internet Versus an Intranet—More Than Just a Typo 9

Birth of the Intranet ... 9
Developmental Differences ... 10
Platforms ... 11
Servers ... 11
Netscape ... 11
Microsoft ... 12
Browsers ... 12
Text-Only Browsers ... 12
Graphic Browsers That Do Not Support Scripting or Java ... 13
Netscape Browsers ... 13
Microsoft Internet Explorer ... 13
Intranet: Development Concerns, Part II ... 14
Predictable Hardware and Software ... 14
Predictable Performance ... 14
Trusted Program Code ... 14
The Least You Need to Know ... 15

3 World Wide Web + Emerging Standards = Chaos 17

Tearing Down the Software Berlin Wall ... 17
A Brief History of HTML and the World Wide Web ... 18
What Is HTML? ... 19
How HTML Works ... 20
HTML Standards and Versions ... 20
Browsers and Scripting Languages ... 21
Web Editors ... 22
The Least You Need to Know ... 22

4 Microsoft Visual InterDev: Order From Chaos 23

Expanding the Set of Web Tools ... 23
Comparing Tools ... 24
Examining the Visual InterDev Tools and Functions ... 25
Sorcery—The Wizards of Visual InterDev ... 25
The Documentation and Research Facilities of Visual InterDev ... 26
Creating New Files in the Project ... 26
Adding Office Documents to a Project ... 26
Visual InterDev Tools Collection ... 27
Database Tools ... 27
Prepare the Easel and Harpsichord—Pictures and Music ... 28
The Least You Need to Know ... 28

Part 2: Understanding the Visual InterDev Interface 29

5 Toolbars/Menus/Help and Documentation: What Does This Do? 31

Documentation and Help 32
Toolbars 34
Just Park It Here—Docking and Floating Toolbars 35
Look Ma, More Toolbars—Showing or Hiding a Toolbar 36
Where Is the Printer Icon? Customizing a Toolbar 37
Creating a New Toolbar 39
Menus 40
The Least You Need to Know 43

6 Creating and Editing Workspaces, Projects, and Files 45

Projects and Workspaces 46
Give Me a Tab 47
Creating a Project 48
Create a New Web? 49
Working Directory and Working Copies 52
Web Directory and Web Copies 53
Adding and Editing Project Files 55
Adding a Page to the Project 55
Editing a Project File 57
Checking the Link 57
Removing a Project 58
Creating the Sample Projects 59
Installing Dos Perros (Two Dogs) Chili Company 59
Installing the 401(k) Sample Application 62
The Least You Need to Know 62

Part 3: Active Server Pages 63

7 Active Server Pages—Don't Just Sit There! 65

Active Server—the Solution to Web Evolution 66
Understanding Microsoft Active Server 67
How Does the Server Know To Be Active? 68
Creating an Active Server Page 68
Creating the Active Server Page 71
Program Logic for the Active Server Page 72
What the Script Does 73
Testing the Active Server Page 73
Active Server Page (ASP) Version 73
The Browser Version 74
The HTML Version 75
The Least You Need to Know 75

8 VBScript: Lights, Camera, Action! 77

Why Use a Scripting Language? 78
VBScript Language Elements 79
VBScript Data Types 79
VBScript Functions 80
VBScript Variables 80
VBScript Constants 81
Control of Program Flow in VBScript 82
VBScript Operators 83

VBScript Procedures 84
Coding Conventions 85
Using Forms with VBScript 86
Examining the Code 89
The Least You Need to Know 90

9 Java, JavaScript, Java Applets: More Than Just a Cup of Coffee 91

Java the Language (and of Course Visual J++) 92
Java Definition in Detail 92
Java Applets 93
JScript: Microsoft's Implementation of JavaScript 94
Writing JScript Code 94
Variables 95
Data Types 95
Operators 96
Program Flow 96
Functions 97
JScript Intrinsic Objects 97
Creating Objects 98
JScript and Active Server 99
The Least You Need to Know 102

10 ActiveX Controls, OLE & VB5CCE: Making Your Own 105

What is ActiveX 105
Inserting an ActiveX Control into a Web Page 106
The Big Test 109
Creating an ActiveX Control With VB 5.0 Control Creation Edition 111
Enter Your Code Please 113
The Least You Need to Know 114

11 Multi-Tier Applications 115

COM, DCOM, LANS, WANS, and Other Confusing Acronyms 116
Distribution for Load Balancing and Application Scaling 117
Language Independence 118
Visual InterDev and Active Server 118
The Least You Need to Know 119

Part 4: Databases 121

12 Relational Databases and SQL: Information, Please! 123

Come Join Me at My Table 124
Normalization 125
First Normal Form 125
Second Normal Form 126
Third Normal Form 126
Constructing the Normalized Database 126
The People Table 127
Phone Number Table 128
The Foreign Key 128
The Job Table 129
Structured Query Language—SQL 129
Select Statement 129
Joins 130
Views 131
Update Statements 131

Delete Statement 132
Insert Statement 133
The Least You Need to Know 133

13 Query Designer, or Instant SQL 135

Types of Queries 135
Connecting to a Database 136
Designing a Select Query 139
Designing the Select Query 142
Designing the Update Query 144
Designing the Insert Query 145
Designing the Delete Query 146
The Least You Need to Know 148

14 Database Wizards, or Instant Databases 149

Creating a New Database 149
Using the New Database Wizard 150
Creating Tables in the New Database 153
Creating a Database Project for an Existing Database 155
Connecting to the Database 155
Opening the Database 158
The Least You Need to Know 158

15 Database Connectivity—Long Distance, Please 159

ODBC—The Universal SQL Translator 159
ODBC Drivers 161
ODNC Data Source Name (DSN) 161
ADO—Using the ODBC Database 164
Testing the Active Server Page 166
The Least You Need to Know 168

Part 5: Integrating Microsoft Visual InterDev 171

16 Working with Link View: Where Is That Page? 173

Links Request and Load Other Files 174
Using Link View 174
The Link View Toolbar 177
Looking at a Web Site 178
The Least You Need to Know 179

17 Working with FrontPage 97: What You See 181

FrontPage Extensions 181
FrontPage Editor (Visual InterDev Version) 182
Sharing a Project Between Visual InterDev and FrontPage 184
The Least You Need to Know 187

18 Source Code Security/Visual SourceSafe: You Changed What? 189

What Source Code Control Should Accomplish 190
Installing Visual SourceSafe 190
Project Security 192
Visual SourceSafe Database 193
The Working Folder 194
Web Projects 194
Shadow Folder 195
Deploying a Project 195
Checking In and Checking Out 196

Checking Hyperlinks 197
Creating a Site Map 198
Showing Project and File History 199
Showing Differences 200
The Least You Need to Know 201

19 Web Site Security: Guarding the Gate 203

Anonymous Access 204
Known User Access 205
Root Directory, Virtual Directories, and Directory Browsing 205
Read and Execute Permission 207
The Least You Need to Know 208

20 Client Scripting—No, You Do It 209

What Can Client Scripts Do? 209
Button Control 210
Reset Control 211
Submit Control 212
Text Control 212
Password Control 213
Radio Button Control 213
Check Box Control 214
Advantages and Problems with Client Scripting 216
Examples of Client Scripting 217
The Least You Need to Know 217

21 HTML Layout Control—Put It Where? 219

Using the Template HTML Layout Wizard 219
Using the Template Page Wizard 220
Looking at Default.HTM 224
Testing the HTML Layout 225
Creating an HTML Layout Control 226
The Least You Need to Know 227

22 Internet Explorer Script Debugger: It did What? 229

Installing and Uninstalling Internet Explorer Script Debugger 230
Starting Internet Explorer Script Debugger 231
Internet Explorer Script Debugger Windows 231
Edit Window 231
Project Explorer 232
Code Window 232
Immediate Window 233
Call Stack Window 233
The Least You Need to Know 234

Part 6: Finishing Touches 235

23 Arts and Crafts: Media Manager, Image Composer, and Music Producer 237

Including Image Composer and Music Producer in the Tools Menu 238
Image Composer 238
Image File Formats 238
Which File Type Do I Use? 239
Understanding Sprites 239
A Palette For Every Taste? 241
Learn More in the Help Tutorial 242

Music Producer 242
Changing the Sound 243
Composing Music 243
Music File Formats 243
Media Manager 244
Where Did My Help Go? 246
The Least You Need to Know 246

24 Creating Your First Sample Project: Where Are the Blueprints? 247

Project Specification 248
Creating the Project 248
Adding the Default.HTM Page 249
Examining the HTML in Default.HTM 250
Adding an Active Server Page 251
More HTML Code 252
Adding a Client Script Page 254
The Script Wizard 256
Connect to a Database 257
HTML Layout 259
Scripting for an HTML Layout Control 260
ActiveX 261
The Relaxation Room 263
Search 263
Check the Links 263
The Least You Need to Know 264

Part 7: Final Issues 265

25 My Web Server's Better Than Your Web Server 267

Administration 268
General Settings for Web Services 269
Directories 270
Editing a Directory 271
Logging 272
Limiting Access by IP Address 273
The Least You Need to Know 274

A Installation and Setup of Software 275

Windows 95 Personal Web Server 276
The Properties Dialog Box 277
Windows NT 4.0 Internet Information Server 278
Microsoft Active Server 278
FrontPage Server Extensions 279
Microsoft Visual InterDev Client 279
Microsoft Image Composer 280
Microsoft Media Manager 280
Microsoft Music Generator 280

B What's on the CD? 281

Legal Stuff 282

C Speak Like a Geek: The Complete Archive 283

Index 293

Introducing Microsoft Visual InterDev

"Today, developers who are on the leading edge are involved in development for the World Wide Web."

"All computer applications are moving to the Web."

"The Web is where it is at."

"You aren't happening until you have a home page on the Web."

These are a few of the countless quotes that you hear everyday. It is true that there is a major paradigm shift towards the World Wide Web. Microsoft, Sun Microsystems, and Netscape are among the organizations busy competing for dominance of the software that will be used for Web server and browser software. Almost every TV ad includes a Web address such as http://www.buymyproduct.com.

The World Wide Web is a cacophony of diverse home pages and content, some of which are boring, some of which can't be mentioned in polite society, and some of which are absolutely marvelous in the richness of design and depth of content. For a home page to be noticed today, it must possess good graphic design, it may need to incorporate sound, it should be interactive, and it must have something to say that the visitor finds worth pausing to investigate. While your home page is only a mouse-click away from anywhere on the Web, the visitor can leave your Web site just as quickly as they came—with just one mouse click.

Good tools are the key to rapid, effective Web development, as developments such as Java and ActiveX continue to stretch the boundaries—and functionality—of the Web. With this in mind, Microsoft has created the first integrated Web development environment.

With Microsoft Visual InterDev, you can use the tools required for all aspects of Web development from creating a simple Web page to creating an elaborate, database-driven, interactive Web site.

In this book, you will explore and gain an acquaintance with the many features of Microsoft InterDev. When you have finished, you won't be an expert, but you will:

- Possess an understanding of the capabilities of Microsoft Visual InterDev.
- Know what areas of the application will be the most useful to your endeavors.
- Be capable of using the various tools.
- Know the areas in which you may want to do further study.

But You'd Have To Be a Genius

Not really. Perfectly ordinary mortals (like you and me) are allowed to tread these hallowed halls. You don't have to know how to write programs. You don't have to know how to write music. You don't have to know how to hack the registry. Your mission is more important, and you'll want to learn fun stuff such as:

- How to install Visual InterDev.
- How to get answers from the online documentation.
- How to write server scripting.
- How to write client scripting.
- How to work with databases.
- How to use the graphic and music tools to make your creations look and sound as though built by an artist.
- How to find all of the various features and functions of Visual InterDev.

In this book you will create and build your own state-of-the-art Web site. You will possess the knowledge to dazzle your friends with high-tech terms like *Java Sandbox*. You will be pleased by the ease with which you will be able to create Web pages using the powerful set of tools in Microsoft Visual InterDev. You will be able to leap tall buildings at a single bound. No, that was Superman. Well, you will at least be able to think about jumping over a foot stool.

And You Are...?

To write this book, I had to envision you, the reader. I needed to listen to you tell me what you wanted to know and what you considered too much detail.

I discovered that there are three of you reading this book. One of you is a savvy computer pro who needs a quick overview of Microsoft Visual InterDev. One of you is a relative novice who wants to be able to create your own Web home page and has decided to use the best tools.

Even if you don't fit one of these profiles, welcome. You will be exploring a marvelous tool and I will do my best to make the trip fun and profitable. Since this is directed at a somewhat diverse audience, there may be parts that you will want to skip since you already know it or you are not interested. Every attempt has been made to make each topic self-supporting. If there is material elsewhere in the book that you may need to understand a particular topic, there will be cross references.

How We Do Things in This Part of the Country

There are several conventions and standards that are used in this book that are intended to make the book easier to use. We hope that it works out that way. Here is a list:

- Any text you need to type or items you need to select appear in **bold**. As an example, you might be directed to click the **Start** button.
- If you need to hold down multiple keys at the same time such as **Shift+Alt+Del** they will be indicated with a plus sign (+) between the keys.
- If you need to press more than one key in succession, they will be separated with a comma. To illustrate, to open the File menu you press **Alt**, **F**.
- If you need to select a short cut key such as **F** for **F**ile where the **F** is underlined, the **F** will be in bold.
- New terms that will be in the Glossary, "Speak Like a Geek," near the end of the book (we try to hide the embarrassing parts as much as possible) will appear in *italics*.

In case you want to know more about the bloody details of Microsoft Visual InterDev, you will find background and in-depth information in boxes. These can be skipped, but in case you are interested, watch for the following icons:

Check This Out

These boxes contain warnings, notes, and other information about Microsoft Visual InterDev. Also included are occasional jokes and other irrelevant attempts at humor.

Techno Talk

These boxes contain high-tech info that is intended to both impress you with the great knowledge possessed by the author and perhaps increase your understanding of the why of some features.

Common Trademark Courtesy

In an attempt to place the blame for everything on someone else and to satisfy an insane desire to avoid law suits, we have determined that we will list the trademarks and service marks for all computer and program manufacturers mentioned in this book here. This way you will know who to hold responsible for the great benefit brought to our lives by these wonderful and generous people, and also for the frustration and aggravation of working with computer software. In addition, if we even remotely suspect a term of being a trademark, we capitalized it.

- Microsoft
- Sun Microsystems
- Intel

Software and Hardware Requirements

Since you have purchased a book about Visual InterDev, you may be interested in knowing what is required to be able to install the software on your computer. Then again, you may not. Just in case, here are the requirements.

Software Requirements

According to Microsoft, Visual InterDev will only work if you have either Microsoft Windows 95 or Microsoft Windows NT 4 installed on your computer. I have accepted their word on this.

Also note that if you are going to use Microsoft Visual InterDev with MS SQL Server, it will need to be MS SQL Server 6.5, and Service Pack 1 for MS SQL Server will need to be installed. The Service Pack comes on the installation CD.

Hardware Requirements

Microsoft provides a minimum configuration and a recommended configuration.

First the minimum configuration. For a Windows 95 installation, 16 MB of ram and an Intel 486/66 class processor. For a Windows NT 4 installation, 24 MB of ram and an Intel 486/66 class processor. My personal recommendation is not to try the minimum configurations unless you are very young and have years to wait for your system to respond. Microsoft minimum system recommendations usually give new meaning to the word "optimism."

The recommended configuration is very realistic (I have tested it and it works just fine). For Windows 95, 32 MB of ram and an Intel Pentium 90 class processor. I have personally used a system with 16 Megs of ram and an Intel Pentium 75 processor, which is a little sluggish but very usable. For Windows NT 4, Microsoft recommends 32 Megs of ram and an Intel Pentium 90 class processor.

Keep in mind that these configurations are for development machines, not production servers. To configure a production server for optimum performance, Voodoo is my preferred method, closely followed by the dart board technique and the SWG (scientific wild guess). In any event, configuring a production server is beyond the scope of this book.

Part 1

What is Microsoft Visual InterDev and How Does It Open Internet Web Technology?

Now is an exciting time to start developing for the World Wide Web. Visions of captivating Web pages dance through your head. You start opening the View Source option on your browser and say, "I can do that!" And indeed you can. It is possible to create a Web page with nothing more than a text editor such as Notepad. The challenge arises when the number of links become more than you can remember, when you begin to use frames, graphics, sound, databases…

In this brief section, you are going to review a short history about the origins of the Internet and World Wide Web; you'll also see where the Internet is going and what tools Micorsoft Visual InterDev gives you.

Chapter 1

Internet Technologies in Perspective: Some History and Facts

In This Chapter

- Review a brief history of the Internet
- Explore the services available on the Internet
- See why TCP/IP works so well for the Internet

The Internet was an outgrowth of a need perceived by the U.S. Military of a redundant and virtually unbreakable communications network. This was back in the 1970s when we were convinced that nuclear attack was a very real possibility. The resulting ARPANET (Advances Research Planning Agency Network) met the military planners' need for a network that would survive the destruction or disabling of any individual piece or pieces. In topology, it was much like the U.S. Interstate Highway system. If you find I-40 blocked between Tulsa and Amarillo, you can detour around through Dallas or Kansas on I-20 or I-70.

ARAPNET was also *packet switched*. Packets are units of data or pieces of files that, on their own, travel on the Internet like cars on a highway. Unlike train cars which travel as a unit, each data packet follows a different route to the destination, and not all packets have to follow the same route. For example, if the first 10 packets of data go down I-40, which then becomes blocked, the next packet or group of packets can follow another route. The packets are assembled into the original file at the other end.

After ARPANET had been around for a while, there were some people who began to see how great this could be for sharing information. The network split into MILNET (for the military) and ARPANET (for the research community). ARPANET then evolved into the Internet, which was used by universities and research institutions to share data. All commercial use was strictly forbidden in the early days of the Internet. Then, the Internet was opened to commercial usage—and a great explosion occurred.

Internet Services

During the early period of development, several services grew up on the Internet. Some of these services are still very important, while others are fading from use. Here is a quick review of the more important Internet services:

- **FTP (File Transfer Protocol)** This is one of the more heavily used services on the Internet. When you visit a site such as microsoft.com and go to the download page to download a file, you are using FTP. Internet sites may have a server dedicated as an FTP site. You connect to an FTP site, locate the file that you need, and download it. If you need to send a file to another computer, FTP can also be used to upload files. Because FTP is so reliable, it would be most surprising to see any thought of replacing it. One of the limitations of an implementation of FTP alone was knowing where the FTP site was located that contained the file that you want to download. The World Wide Web is the current solution to this problem of locating files and sites. FTP is indeed an important piece of a Web site.
- **Gopher** This was one of the better solutions to the accessibility problem presented by FTP, and it is still being used by many educational institutions although its importance is diminishing. Gopher is a protocol that presents a listing of the files at a Gopher site and also has pointers to other Gopher sites. It was developed and is still maintained by the University of Minnesota (thus its name—Minnesota's mascot is the Golden Gopher). Several services that you may hear in conjunction with Gopher are Veronica, Archie, and Jughead.
- **E-Mail (electronic mail)** E-mail, like rock'n'roll, is here to stay. I would even say it is beginning to supplant the U.S. and other countries' postal services (called snail mail). All e-mail uses an *SMTP* protocol (Simple Mail Transport Protocol) to move messages over the Internet to your POP3 (Post Office Protocol 3) mail server where you use an e-mail client, such as Eudora, to read and reply to your mail. In the Microsoft world, e-mail is handled by Microsoft Exchange. Microsoft is working on a new release of Exchange that will embrace the POP3 protocol and tightly integrate with the World Wide Web.

- **Newsgroups** Also called Use Groups, these function much like a bulletin board on which you post e-mail messages. There are over 10,000 news groups covering topics ranging from Alabama to computer programming to zoos. You connect to a news group with your news reader software, subscribe to a news group, read messages and reply to the messages at your leisure. If you are not familiar with news groups, there is a definite set of Internet rules, called *netiquette,* which you should learn and follow. There are two reasons for following the rules. First, it is the polite thing to do. Second, you will avoid being *flamed* by someone who does not have patience for violators of netiquette.

News Groups of Interest

One news group that you may want to subscribe to is Microsoft.public.activex.vinterdev, which is full of comments, questions, and answers from users of Microsoft Visual InterDev.

If you don't have a news reader, you can get one free from Microsoft. Simply use a Web browser to go to **http://www.microsoft.com/ie** and download the latest version of Microsoft Internet Explorer, which will include the Microsoft News Reader client.

- **Telnet** In a Telnet session, you log onto a remote computer and use that computer's resources and programs. Your computer is merely emulating a "dumb" terminal; that is, it is acting like a CRT and keyboard, and not a computer. (One of the common terminals to be emulated is a DEC VT-100.) Say you want to analyze some data using a software package such as *SPSS* (Statistical Package for the Social Sciences). You may not want to purchase the software, so you make arrangements with the local university or college for time on a mini or mainframe through a Telnet session. Telnet is still—and will continue to be—in use.
- **World Wide Web or WWW** Pronouncing this "dub dub dub" (with no rub-a) will definitely establish you as an insider. This is a graphical presentation of information with *hyperlinks*. It was created at CERN in Switzerland as a method of sharing information with graphical content over the Internet. Also included was the idea (borrowed from Gopher) of including links to other Web sites and documents through hyperlinks. Since the balance of this book is about the World Wide Web, no more will be said here.

Most of the services listed previously are beyond the scope of this book. Since they coexist on the Internet with the World Wide Web, you may want to do some further study and experimentation with these services. There are many good books that cover these services in detail (shameless commercial warning—the best books come from QUE Publishing, such as *Using the Internet, Second Edition*, by Jerry Honeycutt).

Transfer Connect Protocol/Internet Protocol (TCP/IP)

TCP/IP is really two separate things that have become so closely associated that they seem like a set of Siamese twins. The necessary element of functioning on the Internet is domain names.

What is a domain? An example of a domain name is microsoft.com. The last part of the domain name such as .com or .edu tells you the type of organization that operates the site but has no intrinsic significance. A university Internet domain name would work just as well if it were enormousstateu.com.

Accurate Names Be careful not to confuse Internet domain names with Windows NT domain names. Although they are somewhat similar in concept, they are very different things.

The last section of an Internet address, such as .com, refers to the domain of the site. Some of the common examples are:

- .com for commercial
- .gov for government
- .edu for educational institution
- .org for a non-profit group
- .uk for the United Kingdom

This gives you some indication of the type of group that operates the Internet site.

TCP (Transfer Connect Protocol)

TCP works by dividing a file into packets that can be transmitted from computer A to computer B. TCP then adds the address of computer B to the packet and shoves the packet out onto the Internet. Various computers on the Internet look at the address and know which path to send the packet through. This is a little like sending a young child on an airplane trip alone; a tag is attached to the child telling where the child is bound for. Various airline personnel look at the tag to make sure that the child is on the correct plane.

If the file is divided into packets, then in my strained analogy, we are dealing with a family of children (that is, packets of information). In order for the destination location to know whether the whole family has arrived, we must include information on the number of children, as in "this is child 1 of 3." TCP doesn't care what the order of arrival of the packets is, but it checks to see if all packets have arrived. If packet 2 does not arrive, the receiving location requests that it be resent.

IP (Internet Protocol)

IP is the addressing function on the Internet. With literally millions of computers and communication line connections making up the Internet, addressing a computer is a key function. The actual address of a computer is a 32-bit address. When this address is converted to decimal numbers, each pair of bytes is separated by a period, as in 127.0.0.1. In Hexadecimal this would be 7F 00 00 01. You will agree that each of these is somewhat difficult to remember. Something like www.microsoft.com, on the other hand, is a little easier for humans to remember.

However, computers don't know what to do with such an "easy" name. To resolve this problem, there exist on the Internet computers that serve as Domain Name Servers (DNS). When you send a request for service from the computer that is the server for www.microsoft.com, this request is sent to a DNS which looks up www.microsoft.com in a list of addresses and finds that the real address is something such as 207.123.211.3. The Internet routers know where 207.123.211.3 is, but www.microsoft.com is meaningless to them. (The address in the hexadecimal format —CF 7B D3 03—is not something that you will usually see when working with the Internet.)

OK, so now you're asking yourself "How does the address 207.123.211.3 get assigned to microsoft.com?" This is accomplished by registering the domain name with a group called *InterNIC*. InterNIC is a cooperative run by the National Science Foundation, Network Solutions, Inc., and AT&T. If you want to learn more about InterNIC go to **http://wwwinternic.net**.

The Least You Need to Know

Understanding the other services and resources on the Internet can greatly enhance your ability to perform effective Web development.

➤ In addition to the World Wide Web, there are many resources on the Internet that you should investigate and become familiar with. Among these are Telnet and FTP. If you want to become adept with the Internet, be prepared to study.

➤ You should know how TCP works in conceptual terms. The relationship between a domain name and the IP address is worth understanding.

- Know how an IP address is resolved. The role of DNS servers is very crucial to the Internet or an intranet.
- Be familiar with the ways InterNIC assists in address resolution and assignment. Understanding how this is done on the Internet will help your thinking if you are working with an intranet.

Chapter 2

Internet Versus an Intranet—More Than Just a Typo

In This Chapter

- The Internet and intranets defined
- Internet and intranet development issues

One of the great strengths of the Internet is its platform independence. This flexibility is a very powerful tool in systems development and integration. Computers and their users are able to communicate with all Internet services regardless of whether they use MS-DOS, PC-DOS, UNIX, LINIX (or one of the other varieties of UNIX), IBM OS/2, or Microsoft Windows (all versions). Internet servers, Web browsers, FTP clients and the rest of the Internet client/server software function on any of these platforms.

When using the Internet, you are usually not aware of the platform that the server is using. For a client system, all that is required is a connection to the Internet by dial-up or dedicated line and the client software (such as Eudora for e-mail or a Web browser for World Wide Web interface).

Birth of the Intranet

Recently, businesses and information systems personnel began to look at the flexibility of Internet technologies, and discovered many advantages over the more traditional client/server technology. Businesses prized Internet technologies because of the following reasons:

- Transitioning to new hardware/software platforms for an enterprise wide information system is much easier.
- They preserve flexibility for the future.
- Database access has been added to Internet technologies, making possible the gathering of information as well as its distribution.

Because of reasons such as this, the intranet was born. Basically, the purpose of the Internet, as with any information system, is the delivery of information to multiple users. An *intranet* is simply a corporate information system, whether on a *LAN* (Local Area Network) or *WAN* (Wide Area Network) that is private in nature and separate from the Internet.

Client/Server Technology

Client/Server was a proprietary, custom-written User Interface software frontend communicating with a proprietary backend, often a database. Elements such as ODBC kept the frontend somewhat independent of the backend. When the User Interface was changed, the changes had to be distributed and installed on all of the client systems. Many times the UI change would be in response to a change in the database, adding new information or removing information. With Internet client/server technologies, the client is the Web browser which has become a flexible element and need not be changed. The changes can all be driven by the backend.

There are different development considerations and strategies that depend on whether you will be using the Internet or an intranet, as you will see in the next two sections of this chapter.

Developmental Differences

The constants that are present in both the Internet and an intranet are very significant. These include the use of TCP/IP, the Web browser, and the Web server paradigm. The differences involve the choice of specific tools and capabilities. Comparisons are discussed in the following sections.

Platforms

If you want to use program code that is transferred to the client system as part of your Web page, it must have been written and compiled for the client platform. (The one exception to this is Java—see Chapter 9 for more details). Let's say that as part of a Web page, you want to send a small program that performs some task on the client system based upon input from the user. Your Web server would have to identify the platform that the client is running on and transfer the proper version of the program. At the present time this type of strategy is not easily workable. I have seen attempts at this but they seem clumsy and are, no doubt, a nightmare to maintain. This means that if you want to have interactive content in your Web pages, the processing must be done on the server. There are several technologies that do just that, such as CGI and Active Server. They are discussed in Chapter 8, "Active Server Pages—Don't Just Sit There!"

The bottom line of the platform issue is that Internet development must be prepared for any platform and any browser. This imposes some limitations, as you will see in the later parts of this section.

Check This Out...

Interactive Content The term "Interactive Content" has taken on a life of its own. At times you expect to hear a drum roll when it is uttered. All it means is that the user can work with the Web page in some way such as entering data, making a choice as to what the page will do next and so on. An example of static or non-interactive content is a Web page that is just for view. The reader's only interaction is to read the material and move on.

Servers

The server software raises the issue of support (or lack thereof) for various Internet technologies. While there are many server packages available, two of the more popular server software package manufacturers are Netscape and Microsoft.

Netscape

Netscape has several Web server packages available, depending on whether you are using Windows NT or UNIX as your operating system. You should be aware that VBScript, ActiveX, and Active Server are currently not supported by Netscape products (although Netscape is very effective in supporting Java). Like all aspects of Internet software, this may have changed subsequent to this writing. If you are going to develop for a Netscape server, Visual InterDev will have to be used carefully, as you'll have to make sure you don't use ActiveX, VBScript, and Active Server. Because of these limitations, I do not recommend developing a Netscape Server with Visual InterDev.

Microsoft

Microsoft's lead product for an Internet Web server is Microsoft Internet Information Server (MS IIS), which is currently only supported on the Microsoft Windows NT operating system. MS IIS supports virtually every technology available, such as Java, JavaScript, ActiveX, VBScript, Active Server, and CGI. In my humble opinion, it is the most universal of the Internet servers available today. Also, Visual InterDev works hand-in-glove with all of the Microsoft server products. Microsoft is also bringing a broad range of Internet server products to the market that will meet most needs. (No, I am not a paid agent of Microsoft. I just happen to find their products very flexible and useful.)

Varieties of Microsoft Servers

One of the potentially confusing aspects of the Microsoft Internet server offerings is their variety and version numbers. MS IIS, the lead product, is currently in Version 2.0. You also hear MS IIS 3.0 discussed. It is not a new version of the MS IIS product. It is the Active Server add-on for MS IIS 2.0. Got all of that? Now I will stir the pot again. MS IIS 3.0 (Active Server) can also be installed on the Peer Web Services product that is the version of MS IIS that runs on Windows NT 4.0 Workstation. OK, we have it all straight. Hold on! I can also install MS IIS 3.0 (Active Server) on the Personal Web Server that runs on Windows 95. If you are not confused now, will you please explain the Theory of Relativity.

Browsers

Just as there are many Internet server products, there are many browsers. Browsers vary in terms of graphic, sound, video, Java, and ActiveX support. The browsers can be divided into four categories, described in the following sections.

Text-Only Browsers

There are Web browsers that are only capable of displaying text. An example is the LYNX browser. It is text-only, text-only, text-only. No, my keyboard did not get stuck in a loop. I am just driving home the point that there are users that have no graphics capability. HTML provides for this by allowing you to add a text line to display in place of your snazzy graphic. Before you start with the attitude that anybody surfing the Web should have graphic capability or not bother, remember, there are users who can't see your graphics even if they download them. Sight-impaired users can, however, read a description or other text substitute for the graphic. (Microsoft has started including the equivalent of closed captioning on some of their training materials that include full motion video. It works for both the hearing-impaired and those without sound cards).

Graphic Browsers That Do Not Support Scripting or Java

An example of a Web browser that does not support client-side scripting or Java is one of the original Web browsers, Mosaic. Mosaic has very good graphics capabilities, but all client ActiveX objects and Java applets will be wasted on a user with Mosaic.

Netscape Browsers

Netscape Navigator is one of the more common Web browsers used by clients. It provides support for graphics, sound, JavaScript, and Java Applets. The primary technology that Netscape Navigator does not support is the Microsoft ActiveX technologies. This may change in the future. As Microsoft and Netscape battle for the hearts, minds, and wallets of the Internet providers and users, the competitive pressures are enormous. The nice part is that we, as consumers, are constantly treated to new and better products at lower prices. Don't you just love this kind of struggle?

> Check This Out...
>
> **What Do I Do if the Browser Doesn't Perform**
>
> A Web page will have varying appearances and capabilities depending on the Web browser with which it is viewed. One technique that is used by many Web developers is to test Web pages with several browsers such as Microsoft Internet Explorer, Netscape Navigator, Mosaic, and Lynx.

Microsoft Internet Explorer

At this time the Microsoft Internet Explorer supports all of the major technologies used on the Web. There are also versions for Windows 3.1 and Windows NT 3.51, as well as an announced UNIX version and one for the Apple Macintosh. Microsoft is extending the support for the ActiveX technologies to these other platforms with the MS Internet Explorer.

This is a somewhat confused picture at times. There are, at the time of this writing, some new functions for JavaScript in the 1.1 version that do not work except on Netscape Navigator. There are also reports that JavaScript is slow on Microsoft Internet Explorer for Windows 3.1. All of this may have changed by the time you read this.

You Can Play Here but Don't Get Out of the Sandbox!

The Java applet runs inside a program called the Java Virtual Machine. The Java Virtual Machine restricts the system resources the Java applet can access to what is called the sandbox. As long as the Java applet stays in the sandbox, it can't have harmful effects. The sandbox is designed into the language. This means that programmers can never make the Java applet access resources outside the sandbox, no matter how hard they try. Never is a long time. This issue has led to an entire industry of digital signing and certification of the source of software.

Intranet: Development Concerns, Part II

When you are working on intranet development, the first thing you will realize is that it is a much more controlled and disciplined environment. It is probable that the software and hardware in use on the intranet has been prescribed by the information systems management of the organization.

Predictable Hardware and Software

The equipment and software in terms of the operating systems and browsers will not be the random variety that is encountered on the Internet. If it is necessary to have the ActiveX support provided by the Microsoft Internet Explorer, it should be possible to require it as a standard on all client systems.

Predictable Performance

One of the limitations imposed on Internet development is the size of graphics included in the Web page. This is a result of the time required to download the graphic file over a 14.4 modem, which is the modem you need to plan for at present. If your intranet will run on a 10Base-T Ethernet with a speed of 10Mbits per second (almost 700 times as fast as the 14.4 modem), graphic size will be less of a concern. The main issue is that you will know the capacities and capabilities of the intranet and can plan and develop accordingly.

Trusted Program Code

Since your intranet will exist in an environment that is known, it can also be tested. ActiveX controls and Java applets that you may include in your Web pages will not produce unexpected and undesirable results, because you will have tested before deployment. Repeat after me, "I will test before deployment. I will test before deployment." You are becoming sleepy.

The Least You Need to Know

Understanding the variable nature of the environment for which you will be performing Web development is a key to success. You have just reviewed several of these issues. As additional technologies come to be used in the Web, additional issues will be added to the list.

- There are issues about the Internet server and browsers you may be working with that you should keep in mind as you develop Web pages. You need to check the feature support for the browsers that you will encounter and how those mesh with the server that you have chosen.
- Internet and intranet development have somewhat different rules. This is because of the differences in the audience, and the security and control requirements of each environment.
- Everything about Internet technology is changing rapidly, so check the status of various issues such as the feature support list for browsers and servers at the time you begin development. Just because it was true yesterday doesn't mean it will be true today.

Chapter 3

World Wide Web + Emerging Standards = Chaos

In This Chapter

- Learn the history of HTML
- Understand the theory of HyperText
- Use HTML tags
- Know the role of W3 Consortium in HTML standards
- Understand the different versions of HTML support by various browsers
- Know the role of browser support for scripting languages
- Realize the role of an HTML editor

The major reason that the Internet technologies—particularly the subset that are the World Wide Web technologies—have become such a focus of development recently is that they offer solutions to problems that have vexed information systems professionals since the beginning of computers. Although we will never find an answer for all of the problems, the Internet technologies are removing some long-standing barriers. The most intractable of these barriers has been the interoperability of software and the moving of applications from computer to computer.

Tearing Down the Software Berlin Wall

The first problem is "How do I get this application software that was created on computer A to work on computer B?" The first solution for this problem was the programming language COBOL (Common Business Oriented Language) that many now feel should be in the Smithsonian. Having written many, many, many lines of COBOL myself, I have a different picture. Each programming language in turn was going to revolutionize the way computers were used. Moving application programs from one computer platform to another is still a daunting task, and expensive.

The second problem is "How do I get computer A to talk to computer B?" The advent of client/server computing has made this an even more pressing problem. The essence of client/server computing is the division of labor between computers. As an example, the databases for an organization are placed on one computer. This computer is connected to other computers over a network. The other computers perform all of the display and manipulation of data. This arrangement enhances and speeds the process. Smaller, less costly computers can be used since the work is distributed.

As you will see in Chapter 12, "Relational Databases and SQL: Information, Please!" there has been a significant amount of standardization in the relational database area. Client/server computing needed an interface, a type of "middle-ware" that would allow the client/user interface software to be independent of which relational database management system is being used on the server to store the data. Microsoft created just such a solution with ODBC (Open Database Connectivity) software.

HTML (HyperText Markup Language) and World Wide Web technologies are beginning to offer a solution to these and other problems.

A Brief History of HTML and the World Wide Web

The term HyperText was first coined by Ted Nelson in 1965 to describe a theoretical model of documents containing non-linear links to other documents, which in turn contain links to other documents. In the theoretical model envisioned by Nelson, all documents that were resident on a computer could be linked to all other documents on the computer. If you were reading a document about the U.S. Civil War and you came to a mention of General Robert E. Lee, you could use the name as a link to another document that might give his history. As you are reading the document about R.E. Lee, you could use a reference to Virginia to go to a document that provided information about the state. There you find the name of the Virginia State Flower, where you linked to a document about botany, and so on. You get the idea.

At the time Nelson proposed the idea, the cost of computing prohibited the use of a computer as an online reading and research device as they are now used.

However, in 1989 a group of scientists at CERN in Switzerland created a method for sharing documents over the Internet. The documents contained HyperLinks (a variation on Ted Nelson's idea) and also contained graphical content. The revolutionary part of this development was that it worked across different computing platforms. Computers of varying operating systems could store and display information without creating a problem. The HyperLinks also could point to a document on another computer using a different operating system.

This all worked because the group at CERN created HTML. HTML is all in ASCII text, which is the same for all computer systems and software. Circa 1994, Mosaic for Microsoft Windows was created by NCSA (National Center for Supercomputing Applications). This rapidly became the first popular Windows Web browser, and led in large part to the initial popularity of the World Wide Web.

What Is HTML?

HyperText Markup Language is a set of instructions in the form of tags that are interpreted by the Web browser to create the display shown in the browser. HTML adheres to the SGML (Standard Generalized Markup Language) guidelines. The HTML page is all ASCII text. Looking at an HTML page in a text editor and in a browser will illustrate how HTML works. The figure below shows an HTML file as seen in a text editor, in this case, Notepad.

Help with HTML An excellent resource for understanding and working with HTML, *Special Edition: Using HTML 3.2*, is included on the CD-ROM at the back of the book—check it out!

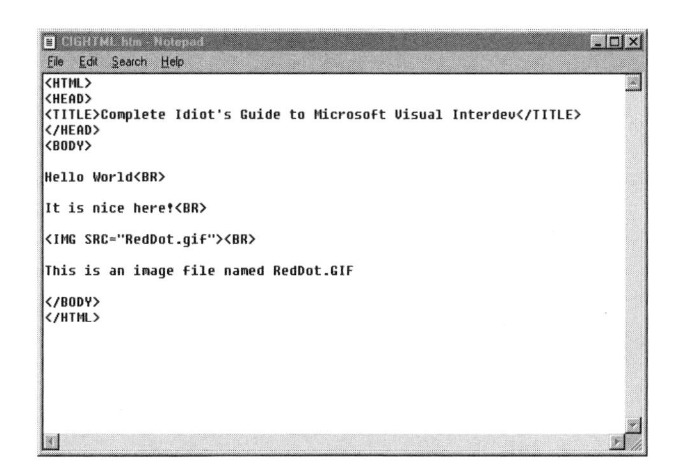

Any text editor may be used to create and edit HTML files.

The result of a browser interpreting the HTML file is shown in the following figure, as displayed by Microsoft Internet Explorer 3.0.

Different browsers will show an HTML file a little differently.

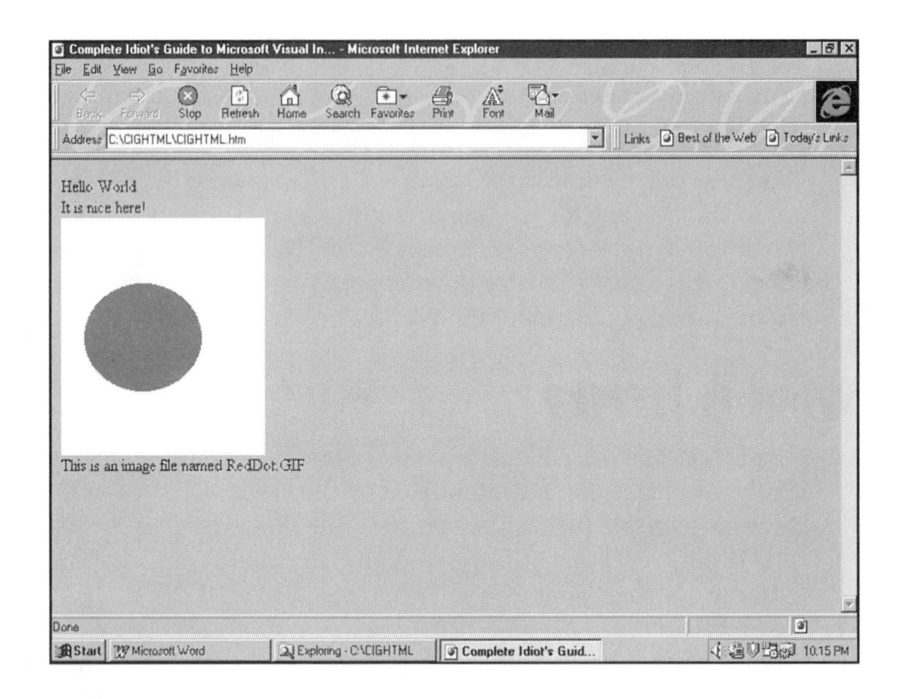

How HTML Works

HTML is text and instructions enclosed in tags. Tags are enclosed in less than and greater than symbols (< >). Many tags come in pairs. An example of a pair of tags is the BODY tags used to indicate the beginning and end of the body of an HTML page or document. These tags are <BODY> at the beginning and </BODY> at the end.

You will get much deeper into HTML as you work with Visual InterDev. The online documentation with Visual InterDev also has a very comprehensive HTML reference section.

HTML Standards and Versions

The current HTML standard in use by the latest browsers and leading edge developers is HTML 3.2. HTML 3.2 is not supported by all Web browsers, such as Mosaic. Mosaic has begun to lag behind as it is supplanted by Netscape Navigator and Microsoft Internet Explorer. Each of the leading Web browser developers has a tendency to create HTML extensions that are supported only by their Web browser. It might be more accurate to say "HTML are..." than "HTML is..." because there are multiple versions.

When you see the statement "This site is optimized for Brand X Web Browser," it means that they are using features of the Brand X browser that are not supported by the Brand Y browser. At present, Microsoft Internet Explorer is the broadest Web browser in the features that it supports. I could present you with a very pretty chart here that compares the features supported by the various Web browsers. However, since you will read this at least a week after I write it, if not more, it is sufficient that you know that everything is changing rapidly. Check what is supported by each browser when you begin development.

What Are the Standards?

"Are there no standards?" you ask (you did ask, didn't you?). Yes, there are. The W3 Consortium is an organization that puts forth standards based upon a consensus of leaders in the World Wide Web technologies. If you want to learn more about this important group, visit their Web site at **http://www.w3.org**. They have a page that gives background information. The URL is **http://www.w3.org/pub/WWW/Press/Backgrounder.HTML**. This will provide a good overview of the role of the W3 Consortium and its important work.

Browsers and Scripting Languages

There are currently two major scripting languages, VBScript and JavaScript. JavaScript is based somewhat on the Java language as developed by Sun Microsystems. However, Sun did not develop JavaScript. It was developed by Netscape in cooperation with Sun. JavaScript is similar to C++ in its syntax. Netscape Navigator and Microsoft Internet Explorer, version 3.0 and later, currently support JavaScript.

JavaScript 1.1 and Internet Explorer

Since Netscape developed JavaScript, the only browser that supports **all** of its features is (not surprisingly) Netscape Navigator. Although other browsers (including Internet Explorer) will work with JavaScript 1.1, it's still best to test your Web pages (all of them, not just those with JavaScript) to ensure they function as you want them to.

VBScript is based upon Microsoft Visual Basic and is very similar in syntax. VBScript is currently only supported by Microsoft Internet Explorer. There is some speculation that Netscape Navigator will eventually support VBScript. As I've stated before, the only constant on the Web is change.

Web Editors

There are at least six important functions of a Web editor or development tool. It should do each of the following:

- Function as a true ASCII test editor. It cannot insert any extraneous characters or formatting symbols as does a word processor.
- Provide assistance in having HTML reference information readily available.
- Provide the means of viewing the page as it will appear in the browser.
- Assist in the editing and creation of graphic content.
- Provide a means of viewing the linkage between pages and other Web sites.
- Provide database and query creation support.

As you'll see in the next chapter (as well as the rest of this book), Visual InterDev fulfills all these criteria.

The Least You Need to Know

The World Wide Web is a diverse place not just because of the variety of content. It is also diverse in the types of usage and the software platforms for both the server and client sides of usage. While it's true that you can perform Web development with very simple tools, you can also build a house with a hammer and a hand saw. It just isn't the way it is done today. Just remember:

- It is important to clearly define the audience to which you will aim your Web site and the browsers that will be used.
- Test with the variety of browser software that you expect to encounter.
- Choose items such as scripting language based upon your audience, not your preferences.
- Choose the tools, languages, and techniques that you will use for Web development as carefully as you will choose your content.
- Everything on the World Wide Web is changing rapidly. Remaining agile is essential.

Chapter 4

Microsoft Visual InterDev: Order From Chaos

In This Chapter

- Review some of the many diverse tools that are pulled together in Visual InterDev
- Examine how the one-source availability of these tools will save time and eliminate confusion

The introduction of Microsoft Visual InterDev is a breath of fresh air in the fetid realm of disorganization. (Who says that you can't leap to new heights of hyperbole in computer books?) Seriously, Visual InterDev is the first comprehensive toolkit for Web development.

Expanding the Set of Web Tools

The expansion of the use of the World Wide Web was a hallmark phenomenon of the early 1990s. Along with the increase in numbers of users came a broadening of the types of use, which lead in turn to an increase in capabilities and tools for utilizing the power of the Web.

Microsoft appears to be responding to this pressure in a very positive manner. The Microsoft Web server software is, at the present time, the most comprehensive in the scope of features and compatibility. Most Web sites, regardless of the platform on which they were developed, will move to a Microsoft server more easily than the other direction. Microsoft Internet Explorer is the most capable browser available today. This is true at least in terms of the capability to display Web pages (as they were created) that include JavaScript, Java applets, VBScript, ActiveX controls, and so on. Microsoft seems to be taking the approach that their software should contain a "super set" of features.

Comparing Tools

Let's examine a few of the steps involved in the creation of a Web site and the different tools that are required. Let's see how many of them Visual InterDev will handle.

Old Tools Versus InterDev Tools

Task	Old Tool	Visual InterDev
Create folder for Web	Windows Explorer File or New, Folder	Created as part of Project Creation
Setup Web folder in Server with alias	Server Administrator	Created as part of Project Creation
Create a Default.HTM file for Web site	Notepad, HotDog or other editor program	Choose File, New, Files, HTML page
Check syntax for VBScript	Find a manual or open documentation	Check InfoView VBScript reference in Workspace
Check HTML tag syntax	Find a manual or open a help file	Check InfoView HTML reference in Workspace
Check Web page appearance in browser	Save and close HTML file and open browser	Right-click HTML page edit window and choose preview
Create new SQL Server database	Open SQL Server Enterprise Manager	Use Visual InterDev Database Wizard
Design queries for use with Web database	Find SQL manual and write SQL statements	Use Visual InterDev Query Designer
Add, modify, or delete data in database	Write ISQL for SQL Server or create SQL script	Open Table in DataView and add data

Examining the Visual InterDev Tools and Functions

When you open a Web project in Microsoft Visual InterDev, you have a complete set of tools at your disposal that allow you to perform the edits, additions, and deletions that are required in the creation and maintenance of your Web project—no longer will you have to jump from tool to tool.

Sorcery—The Wizards of Visual InterDev

Visual InterDev Wizards perform tedious and repetitive tasks that are error-prone because of the strict requirements for correct syntax and form. These Wizards will save you hours of work and frustrating debugging.

- **Database Project Wizard** This wizard creates the files and connections necessary to add a database to a project.
- **Departmental Site Wizard** A wizard that creates a departmental site complete with Web pages that only need to be customized to fit your organization. All of the linkages are already there.
- **Sample Application Wizard** This assists in the installation of sample Web applications and Web applications with an INF installation script file.
- **New Database Wizard** If you are creating a new SQL Server database, there is no need to open the SQL Server Enterprise Manager to create a Database Device, create a database, create tables, or add data to the tables. It can all be done from Visual InterDev.
- **Web Project Wizard** Use this wizard for the creation of a simple Web page.
- **Data Form Wizard** Need an HTML form to add, update, and delete data from a database? Look no further. This wizard will make the process simple.
- **Template Wizard** Want to create some standard layout features and elements for your Web page? Turn here for comprehensive assistance.
- **HTML Insertion Wizard** This wizard generates HTML for the creation of conditional (True|False) logic and Data Range HTML (return and display 10 or 15 records per page).
- **Script Wizard** Want help with the creation of VBScript in your HTML page? Ask the Script Wizard.

The Documentation and Research Facilities of Visual InterDev

One of the great conveniences of Visual InterDev is the fact that all of the documentation and research information that you may require is always available in the Workspace window.

- **Documentation** The InfoView tab of the Workspace window provides a searchable interface to a comprehensive set of on-line reference material. Included are answers to questions about VBScript, HTML, Database tools, Music Generator, and so on.

Creating New Files in the Project

When you need to add a file to the project upon which you are working, choose **File, New** and click the **File** tab to create a variety of new file types. Each file will have several standard items already in the file for your convenience.

- **Text File** You can create a plain generic text file without opening Notepad.
- **Active Server Page** When you need to add an Active Server Page, a new file with the .ASP extension is created with all of the standard HTML code already included.
- **HTML Page** New files with the standard HTML elements and the .HTM extension are created with a few clicks of the mouse.
- **ODBC Script File** When a script is needed for the creation or management of SQL Server objects, you can create and manage the .SQL files from here.
- **HTML Layout** Your .ALX files can be created, edited, and managed here.
- **Macro File** Macros recorded and saved. All .DSM files are welcome.

Adding Office Documents to a Project

When you need to add Microsoft Office documents to your projects, they can be created and edited from within Visual InterDev.

- **Word Document** When there is a need to create or edit a Microsoft Word Document, you can do it from here if you have Word installed on your system.
- **Excel Chart** Excel charts can be created and edited from Visual InterDev. All of the Excel editing features and functions that are available with your version of Excel are available.

- **Excel Worksheet** Want an Excel worksheet? No problem. There are full editing and creation capabilities here. Your .XLS files will work just as though they were running in Excel, because they are.
- **PowerPoint Presentation** Create and edit PowerPoint (.PPT) files with full capability.

Visual InterDev Tools Collection

There are a large variety of tools that have been offered by Microsoft during several months of frantic development. Several of them have been pulled together in Visual InterDev.

- **ActiveX Control Insertion** Insertion of ActiveX controls was at first a somewhat complex process involving the copying of the CLSID. Then came the Microsoft ActiveX Control Pad that simplified the process. Now the Control Pad functionality is available from within Visual InterDev.
- **LinkView** Checking internal links (within the Web) and external links (to another Web site) is one of the more time-consuming and frustrating tasks in the creation of a Web site. Visual InterDev provides LinkView, which creates a visual representation of the Web and checks that the links work.
- **Preview Web Page** Never again will you need to close and save an HTML file, then open it in the browser to get a WYSIWYG view of your creation. Right-click on the page in the edit window and choose **Preview** from the menu to accomplish the job.
- **Properties** You can view the properties of any file in the workspace.
- **FrontPage 97** Projects can be jointly edited and worked on by users of both Visual InterDev and FrontPage. Also, the FrontPage WYSIWYG editor can be used from Visual InterDev.

Database Tools

One of the more daunting tasks is working with databases. The very term "database" strikes fear into the hearts of brave and fearless explorers of computer realms. Visual InterDev provides several guides through these tortuous paths in the form of several database tools. Take heart. Travel these paths, and you will survive and thrive.

- **Query Designer** Is the writing of good syntactically correct SLQ statements giving you a problem? You signed on to develop Web content, not become a database guru. The Query designer is a graphical tool that will create correct SQL every time.

- **Database Access** Need to add data to your database? Want to create a new table in your SQL Server database? Need to update data in your database? Need to add a column to your database table? You can do all of these and more from Visual InterDev.

Prepare the Easel and Harpsichord—Pictures and Music

Multimedia files are becoming commonplace on the Web. Visual InterDev provides excellent tools for creation, editing, and cataloging these files.

- **Media Manager** Organizing, annotating, and searching for multimedia files that you need to use in your Web page is simplified with Microsoft Media Manager. You view and search on annotations for files.
- **Image Composer** This full capability graphic tool will assist you in the creation of graphics for your Web site.
- **Music Generator** Add royalty-free music to your site. Music Generator will not win you a Grammy, however. (At least we haven't heard of such an instance.)

The Least You Need to Know

Many powerful tools are very complex to use. Visual InterDev is an exception. Its simplicity may deceive you into thinking that it can't be powerful. This is the third tool that Microsoft has created on the Developer Studio model. First came Visual C++, followed by Visual J++, and now Visual InterDev. Each iteration of this format gets stronger and more intuitive. If you need something for Web development, it is probably here. Just remember:

- Look in the documentation. The search capability in the InfoView makes the information very accessible.
- Right-click on various objects. You will uncover powerful capabilities that are just a click away.

Part 2

Understanding the Visual InterDev Interface

You are about to embark upon the actual use of the Visual InterDev integrated development environment. The basic idea behind a development environment is that you can invoke any tool to work on the object of your creation. In this case you are developing Web pages which you will put together to create a Web site.

Hold it. Are you going to need editors and tools for a Web page? Yes. And that is where Microsoft Visual InterDev enters the picture. Microsoft Visual InterDev is an interface and organizer for all of these tools. It is a workbench that holds your project while you apply the correct tool.

In the last chapter in this section, you will begin creating your first sample project. Before you do that, you will review the tools in your toolbox to see how each works. This will enable you to see the nuances of the tools while you use them.

The greatest thing about these tools is that you can't smash your finger with any of them. Believe me, if it was possible I would have.

THE PROBLEM WITH A HOME OFFICE.

Chapter 5

Toolbars/Menus/ Help and Documentation: What Does This Do?

In This Chapter

- Find information in the online documentation and Help
- Dock or float a toolbar
- Show or hide the available toolbars
- Show ToolTips and Shortcut Keys
- Customize a toolbar
- Create a new toolbar
- Customize the Tools menu

Having the right tool or function at the right time makes Web projects faster and easier and therefore more fun. Having fun *is* one of the noble purposes of all work, right? Understanding the use and management of the toolbars and menus in Visual InterDev is easy and will keep you from rummaging in a cluttered toolbox for the needed tool. You will be like the surgeon on a TV medical show—when you want a tool, it is there, delivered by a trustworthy, skilled assistant. WithVisual InterDev, the toolbars and menus are your trustworthy assistants. You just need to learn how to ask for the tools.

Documentation and Help

Microsoft Visual InterDev has a plethora of information available for your use as you create and edit your Web site and pages. When you open Microsoft Visual InterDev, it should look much like the following figure. If the Output window isn't open, open it by choosing **View** from the menu bar and choosing **Output**. You will see that the Output window has several tabs along the bottom, as shown in the figure.

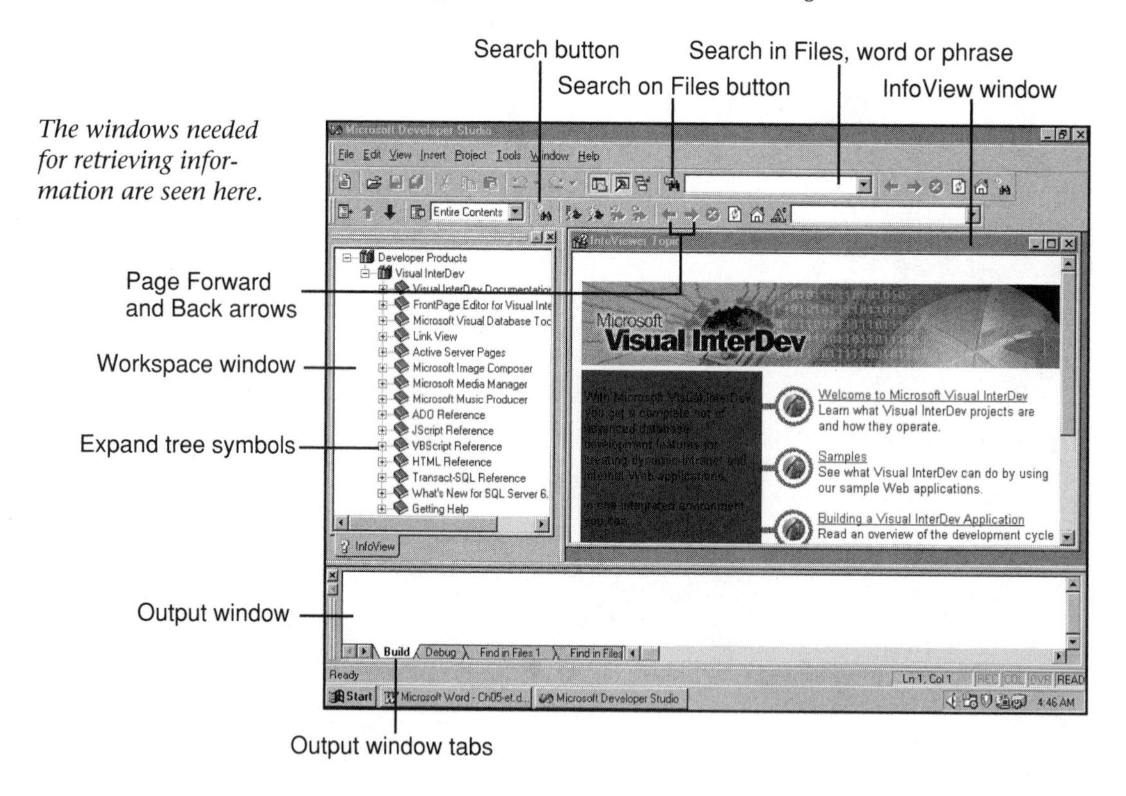

The windows needed for retrieving information are seen here.

An excellent place to start using Help is to click one of the Search buttons. A dialog box opens, asking you for some search criteria.

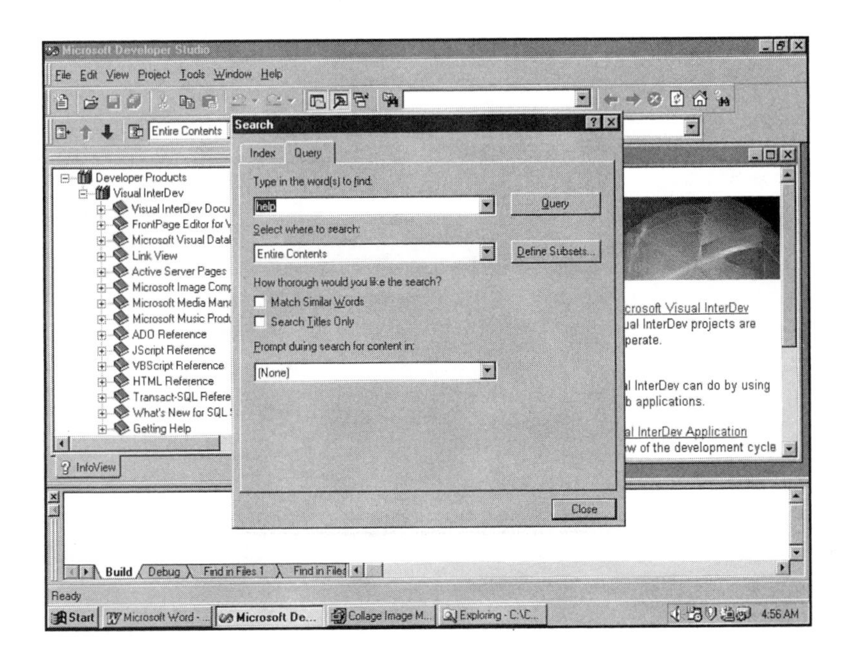

This dialog box will be used to search for information.

Click the **Query** tab and enter "**Help**" in the text box labeled "**Type in the word(s) to find:**" and click the **Query** button. This opens a new window titled Results List as shown in the figure below.

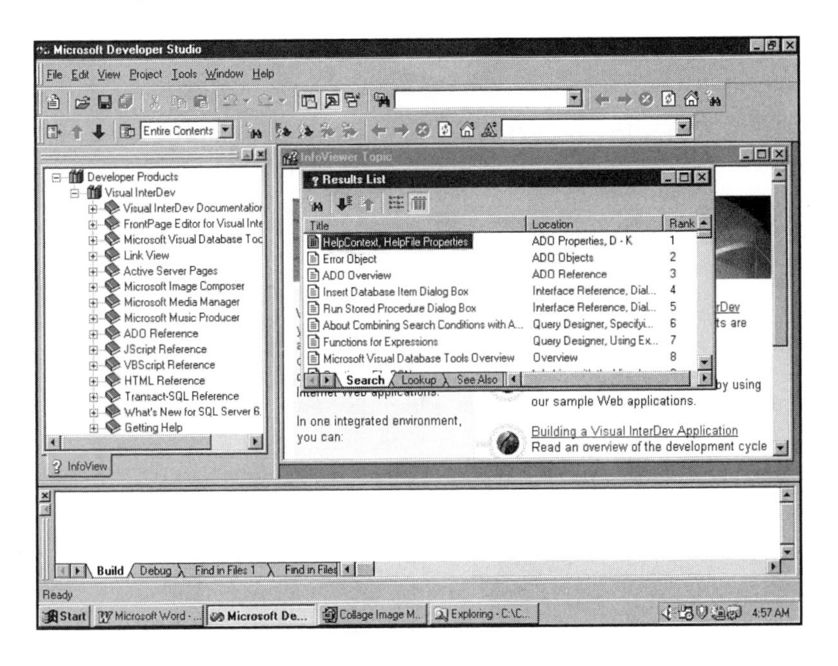

The Results List window contains the list of files that contain the search word or phrase.

If you choose one of the entries in the Results List and double-click it, the topic is displayed in the InfoViewer Topic window.

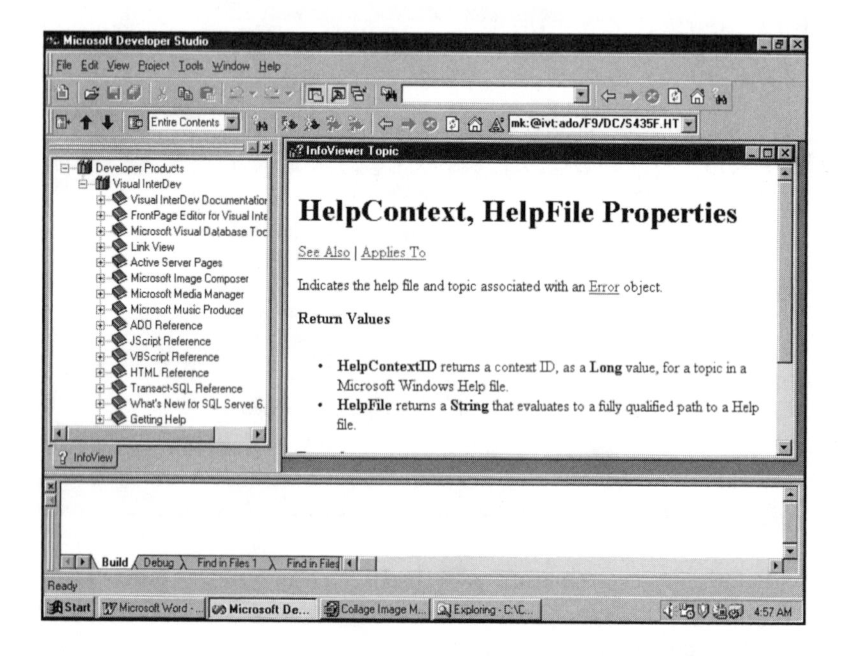

You can switch between the Results List and InfoViewer Topic window as needed.

The Index tab provides a slightly different approach to finding information in the same Help files. Also, choosing Help in the Menu bar will bring you into the same help functions.

The second Help function is the "In Context Help." You access this by pressing the **F1** key when you are editing the HTML of a Web page in the InfoViewer Topic window. The context selected is for the word where the cursor is located. (You will work with this and the Find In Files search in the next chapter.)

Toolbars

You can move toolbars, hide them, and even create new ones that contain just the tools you need. You can "float" or "dock" a toolbar, as well as add tools to an existing toolbar. You can have the toolbars show ToolTips when you move your mouse over a tool icon, and you can make the tool icons large or small. In short, you can probably get a Ph.D. in toolbars, but who wants it? (A master's degree will probably suffice.)

There are six standard toolbars available (note that only two are available—"Standard" and "InfoViewer"— when you first open Visual InterDev):

- Menu Bar
- Standard

- "Visual InterDev"
- InfoViewer
- Edit
- Database

Just Park It Here—Docking and Floating Toolbars

The first time you open Visual InterDev, you should see one or two toolbars near the top of the window. These are the "Standard" toolbar and the "InfoViewer" toolbar. You can dock and float these (as well as other) toolbars:

- **To dock a toolbar** means that you have parked the toolbar on the edge of the window.
- **To float a toolbar** means that it stays on top of the window and you can move it around. When you move a toolbar to the edge of a window, it docks itself there.

Practice is the best way to understand how simple docking, floating, and moving a toolbar really is. A "docked" toolbar has a handle that you can grab with your mouse (a floating toolbar's handle is its title bar). To grab the handle, put the tip of your mouse on the handle, press and hold the left mouse button down and drag the toolbar where you want it and release the mouse button.

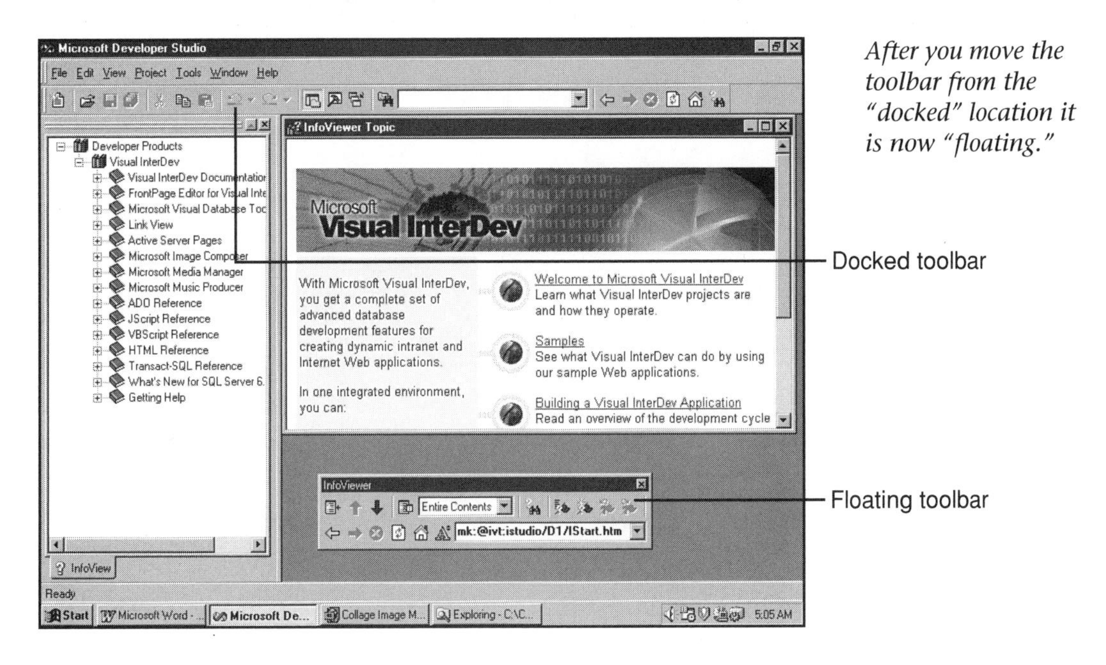

After you move the toolbar from the "docked" location it is now "floating."

A docked toolbar does not cover any of the work in the main window. The window is resized to allow for the docked toolbar while a floating toolbar always stays on top of the work in the main window. (You never accidentally bury the toolbar under other windows.) The great part of this flexibility is that you can park a toolbar where you want it; you can dock a toolbar on any edge of the main window, as well as "stack" toolbars.

Look Ma, More Toolbars—Showing or Hiding a Toolbar

By now you are undoubtedly eager to see the other toolbars that are available. They only seem shy—once you know how to coax them into the open, they will appear at your whim.

When you have Visual InterDev open, choose **Tools, Customize** at the top of the window. The Customize dialog box appears with several tabs at the top. Choose the **Toolbars** tab.

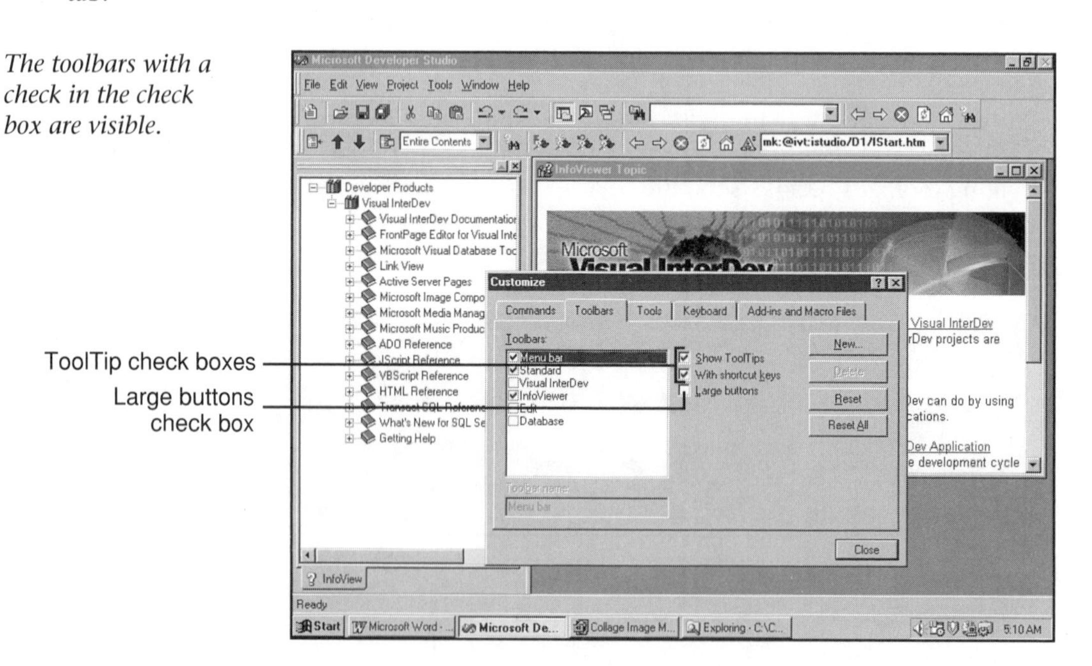

The toolbars with a check in the check box are visible.

ToolTips

By now you may be convinced that you will never remember the meaning of all of these cute little icons. Not to worry. ToolTips to the rescue.

To see a ToolTip, simply place your mouse over a tool icon on a toolbar. A small text box appears that describes the purpose of the tool or displays the tool's name. The box will also list any shortcut keys for the tool (such as Ctl+O for Open, used to show the dialog box for opening a file).

You can turn ToolTips on and off as well as decide whether to show the shortcut keys. Simply choose the **Tools, Customize** item at the top of the window. With the **Customize** dialog box open, choose the **Toolbars** tab. Click **Show ToolTips** to activate this option.

Now you can hide the toolbar by removing the check in the check box by the toolbar name in the dialog box. If you check the toolbar again, it will appear where it was before you hid it. Note that the only toolbar you cannot hide is the Menu Bar. If you hid the Menu Bar, you wouldn't be able to access the Customize dialog box to make the Menu Bar visible again. This is a good safety precaution that the software designers built in to protect you from a very troublesome mistake.

Finally, before you close the dialog box, notice a button on the right side of the dialog box named **Reset All**. This is a useful button when you are experimenting and learning to use toolbars in Visual InterDev. Clicking this button will remove all changes that you have made to the toolbars and put them in factory- fresh condition. So experiment away. You can't hurt anything. When you're finished, close the dialog box by clicking the **Close** button on the bottom-right.

Check This Out...

Large Buttons
When you have the Customize dialog box open, click the **Large Buttons** check box and see what happens to the size of the icons in the toolbars. Cool, huh? Just clear the check box to return to small tool buttons.

Where Is the Printer Icon? Customizing a Toolbar

With all of the toolbar flexibility you have seen so far, you are no doubt wondering if you can add a tool to a toolbar. (If you aren't, you should be. Humor me.) In fact, there are dozens of tools that you can add to the toolbar of your choice.

For practice, you will add a printer tool to a toolbar. Choose **Tools, Customize** from the Menu Bar. Choose the **Commands** tab on the **Customize** dialog box.

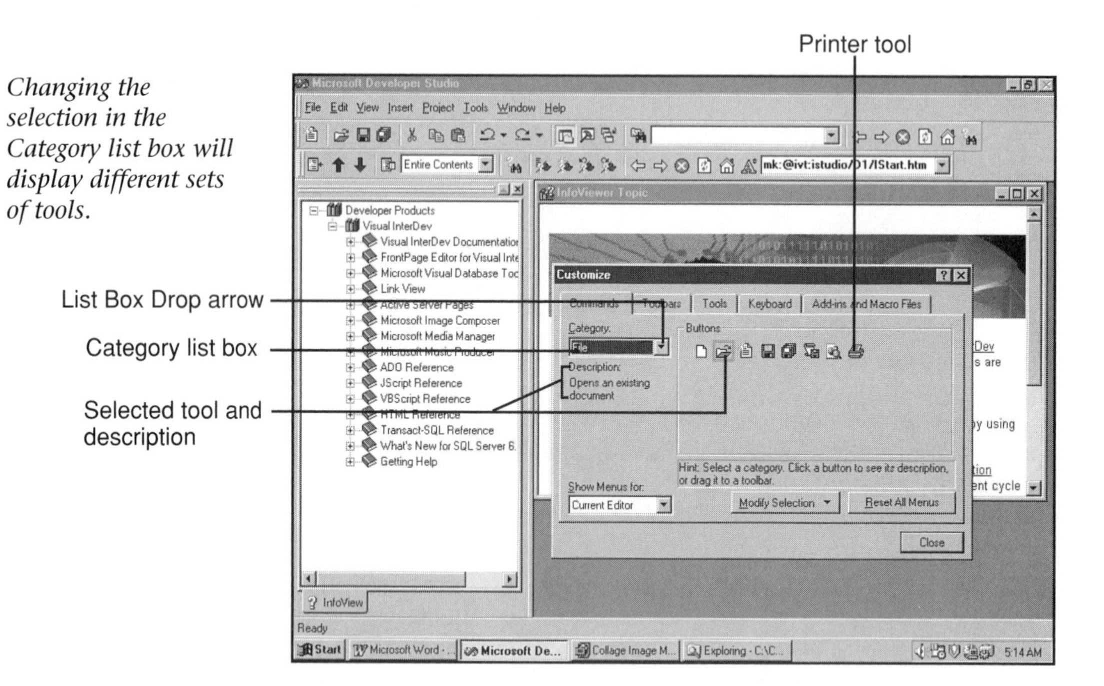

Changing the selection in the Category list box will display different sets of tools.

Now you will add a printer tool to an existing toolbar by grabbing the printer tool with your mouse. Place the mouse over the printer tool. Hold down the left mouse button and drag the printer tool to a toolbar and release the mouse.

To remove the tool from the toolbar, simply reverse the process, grab the tool with the mouse and drag it back to the dialog box.

Tools

In the Command dialog box, you will see a list box titled Category. This is a drop-down list box. Click the down arrow at the right of the box with your mouse. You will now see a list of selections. Select **Edit** and click. The dialog box displays a new set of tools in the area to the right of the list box.

You can see what a tool is by clicking it with your mouse. The tool is described in the description area beneath the Category list box.

If you have time, you may want to look at the other entries in the Category list to see the tools available for your selection.

Creating a New Toolbar

OK, OK, so none of the existing toolbars are exactly what you want. In this case, you may create your own toolbar. The cool part is that it is so easy. The hardest part is giving the toolbar a meaningful name. Just to keep it simple, you will name your toolbar "My Toolbar" for this exercise. (OK, so it isn't original. You may use any name that you wish. You may also use blanks and numbers in the name. Just substitute your choice of names for My Toolbar.)

Start by choosing **Tools**, **Customize** from the Menu Bar. Now, click the **Toolbars** tab. Click the **New** button and enter the name **My Toolbar** in the text box that appears. Click **OK** to accept the name. A new toolbar with no tools will appear.

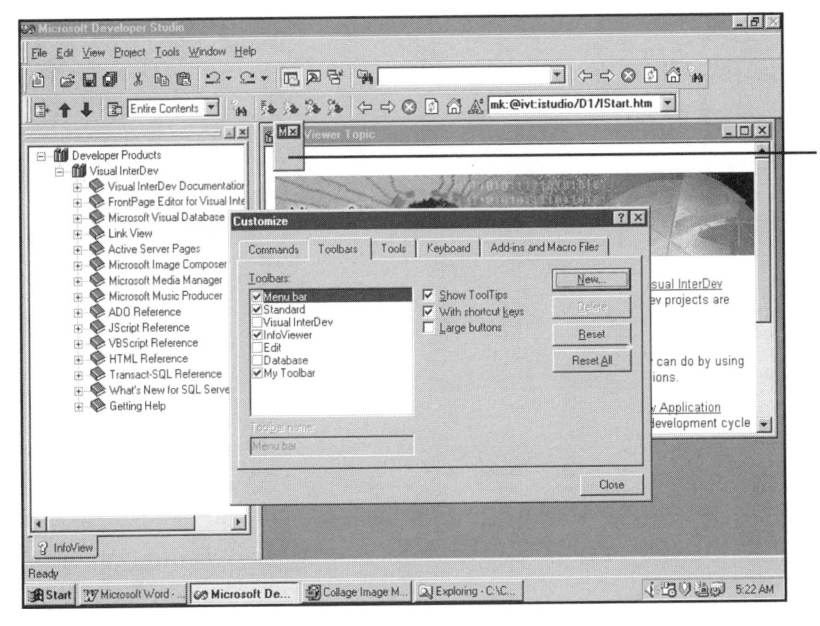

You need to add tools to your new toolbar.

You now have a toolbar that has no tools. Since a toolbar with no tools is not particularly useful, you will now add tools. Click the **Commands** tab on the **Customize** dialog box. Now, start dragging tools to the new toolbar as you did in the previous section of this chapter. Grab each tool with the mouse by holding down the left mouse button and moving the tool to "My Toolbar." For practice, move four or five tools to the toolbar. My Toolbar should now look something like the following figure.

> Check This Out...
>
> **Choosing Tools** You can select tools from any category in the Commands tab. You will want to choose the most frequently used tools. It is acceptable to mix tools from various categories.

You can delete your new toolbar by selecting it in **Toolbars** tab of the **Customize** dialog box and clicking the Delete button. Don't be concerned that you will delete one of the standard toolbars. When you select one of the standard toolbars, you will see that the **Delete** button is grayed out. This means that it is not enabled. Again, experiment. You can't hurt anything!

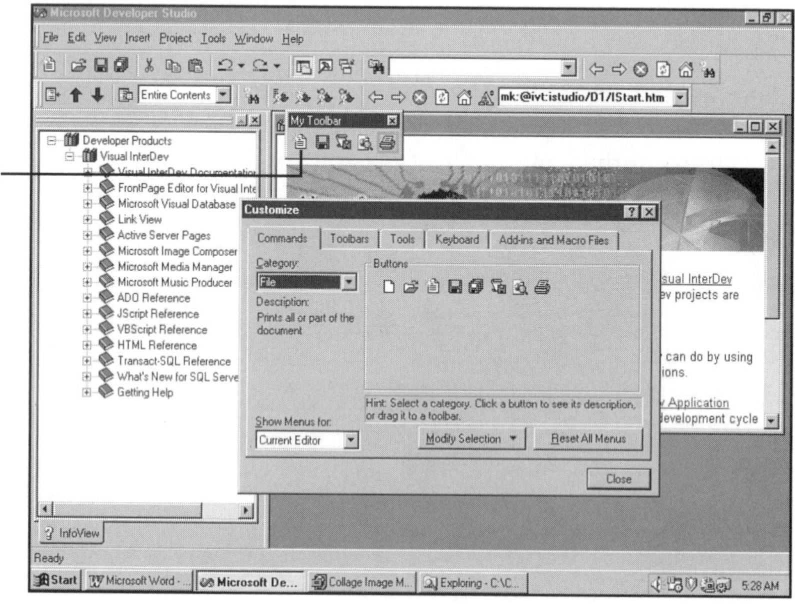

You have populated your new toolbar, My Toolbar, with tools.

Menus

You can customize the Tools menu by adding tools. Examples of tools that you may want to add are the Microsoft Image Composer and the Microsoft Music Producer. Adding them to the Tools menu will allow you to open them without hunting the Icon on your Desktop or in the Start menu. This can be very useful when you are in the middle of a complex edit and you want to check something without minimizing multiple windows.

> **Check This Out...**
>
> **Look and Listen** In Chapter 23, "Arts and Crafts: Media Manager, Image Composer, and Music Producer," you will explore these other tools in more detail. You will create your own music even if you have a tin ear like me.

Start by selecting **Tools** from the Menu Bar and choosing **Customize**. Now choose the **Tools** tab on the **Customize** dialog box. Click the **New** button on the dialog box as shown in the following figure, and you will see the dialog box ready for a new entry.

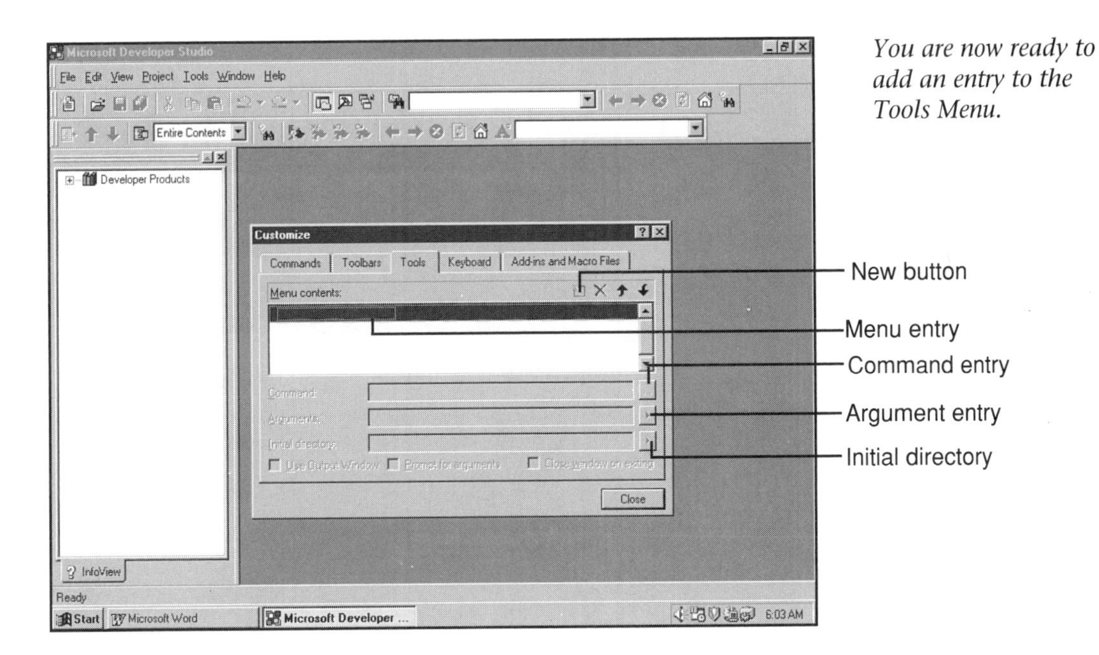

You are now ready to add an entry to the Tools Menu.

You will enter the description that you wish to appear in the Tools menu for the Image Composer. You also need to enter the command. For the Image Composer it is C:\Program Files\Microsoft Image Composer\Imgcomp.exe. This is the path to the executable program. (If you click the arrow button on the right of the command text box, a browse menu will appear that you can use to find the program on your computer.) Next, you can enter the initial directory or working directory. In the following figure the initial directory has been set to be the current directory. This is the default save location for files. It is also the default location that is searched when you attempt to open files.

Techno Talk

Paths and Directories In Windows, for each program or application, there is always a Current Directory. By definition, there must be a current path or directory. For most programs, you can have a Working Directory. The working directory is where the program will attempt to save files, look for files, and so on. The working directory says to the program "do it here." If you set the Working directory parameter to the Current Directory, whatever directory is open at the time is the "do it here" directory.

When you click the Close button, the task is complete.

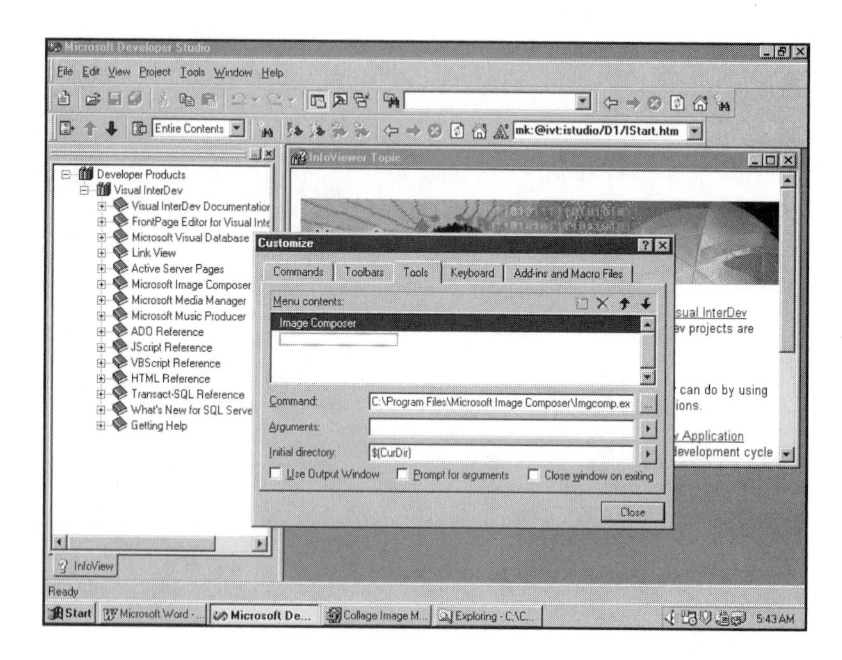

Now when you open the Tools menu you will see a new item.

Click the new Image Composer entry to open this program.

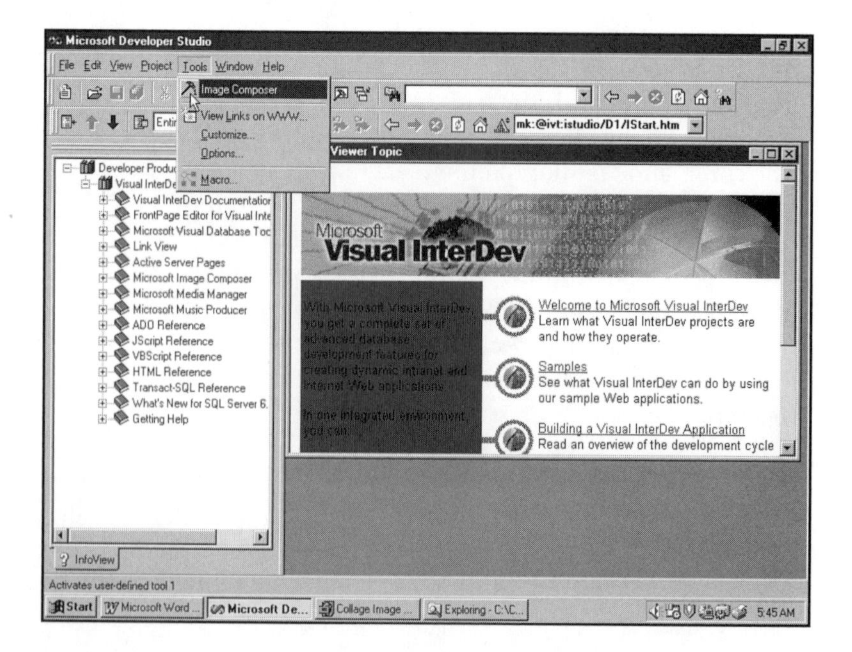

If you want to delete a tool that you have added to the **Tools** menu, you simply open the Customize selection to the **Tools** tab as before, highlight the entry that you want to delete, press the **Delete** key on your keyboard, and click the **Close** button.

Guess what? Experiment all you want. You can't hurt anything!

The Least You Need to Know

You don't have to be an expert to use toolbars and menus efficiently and effectively. If you worked your way through all of the exercises in this chapter, you are entitled to at least a Masters degree in toolbars, (because *you* are now the master, *not* the toolbars). These tools are here for your use in editing and testing your developed pages and Web site. The documentation and help will provide you with needed information and guidance. Just remember:

- The online Documentation and Help is there for you, so use it!
- Toolbars can be relocated where you want them.
- You can show the toolbars of your choice.
- A ToolTip can identify a tool icon for you.
- You can add and delete tools from a toolbar.
- You can create your own toolbar.
- You are in charge, not the toolbars.
- Menus can be customized to fit your needs.

Chapter 6

Creating and Editing Workspaces, Projects, and Files

In This Chapter

- Create projects and workspaces
- Create files in projects
- Learn to add existing files to projects
- Edit files in projectl
- See what happens when a new project is created
- Examine the sample projects supplied by Microsoft
- Learn how to remove an unwanted project and workspace

In Integrated Development Environments for programming languages, the thing that you are working on is a program or application, but in Visual InterDev, the paradigm is different since you are working on a Web. The concepts of the workspace and project as implemented in Visual InterDev are very functional and useful.

Check out the CD-ROM!

This chapter has corresponding sample files on the included CD-ROM. Simply click on "Examples," then the corresponding chapter number for which you are interested. (Be sure to read the Readme text file, also located in the "Examples" section, for important information on installing the sample files.)

Projects and Workspaces

Central to understanding and using Visual InterDev is familiarity with the concepts of the project workspace and the project. The short definition of a project is the set of files required to publish a single Web; there are usually dependencies and relationships between these files, such as links from file to file.

Workspaces contain projects. The term workspace, as it will be used throughout the remainder of the book, has two meanings. These meanings are sort of the same, but aren't. (OK, I'll quit being confusing.)

- The first meaning for workspace is the lower left window that is displayed by Visual InterDev. This window is control central for the files that you are working with and the reference information that you may need. When you first open Visual InterDev, InfoView is the only tab in the workspace window. This tab displays the documentation and reference material as is shown in the workspace window in the followng figure. This window is also where we display a workspace as described in the second meaning.

Clicking the + symbols will expand the list in the workspace window.

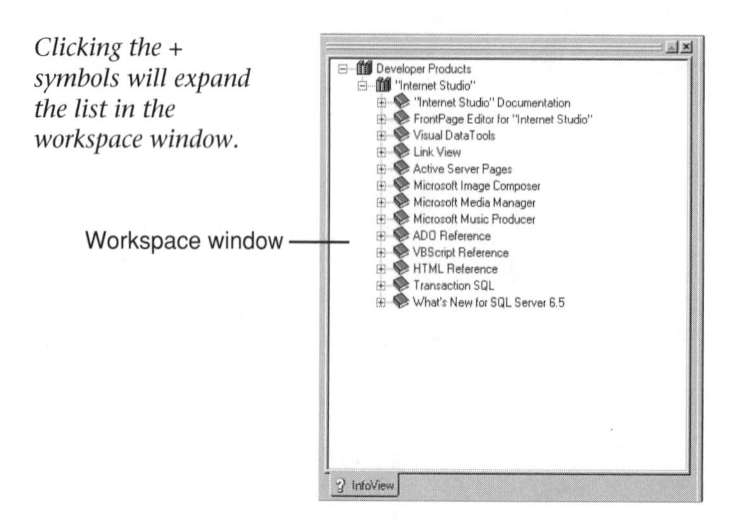

➤ The second meaning for workspace is a bit abstract. It is the set of projects and files that you want to work on at one time. You can name a workspace, and a file is created that will record what you have decided to include in the workspace. If you work in the workspace for a while and then you move to another task, you can reopen right where you left off. It is like having a separate workshop for each project. If you leave workshop A to go to workshop B for a while, when you return to workshop A, everything will be as you left it.

This second definition of workspace is the one that we are concerned with in Visual InterDev. This workspace will contain all of the projects and files that we are working on in that "workshop" and all of the database connections that we need to work with the projects.

Give Me a Tab

When you open a workspace file, at least one (and usually two) or more tabs appear, the DataView tab and the FileView tab. When you click the FileView tab, you see the projects and files that comprise the projects (displayed as in the figure below). Double-clicking one of the file names will open it for editing.

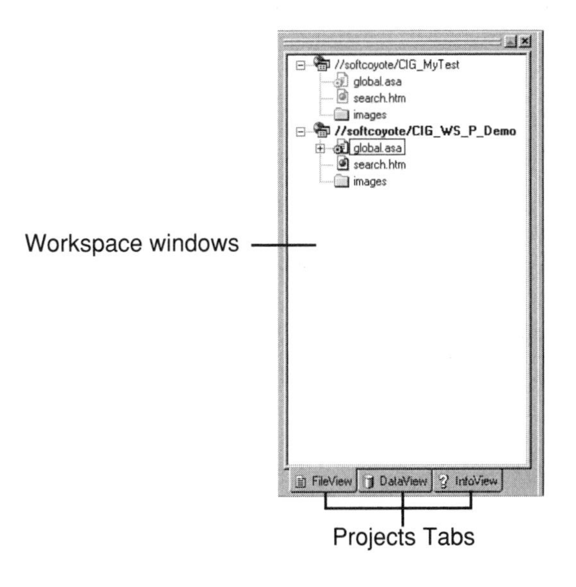

A workspace can contain more than one project. The projects may also be on different servers.

If you included a database connection in the project, the DataView tab is also displayed. When you click this tab, the database objects display in the workspace window. Double-click a table name and the contents of the table display in a grid.

You can create workspaces empty (with no project) or in the process of creating a project. Projects may be added to existing workspaces and files may be added to existing projects. Are you confused? I am. The best way to remove the fog is to actually create a project from scratch. When you watch the process all these different relationships become clear.

If you have sufficient rights, you can add and delete data from tables through the Visual InterDev workspace.

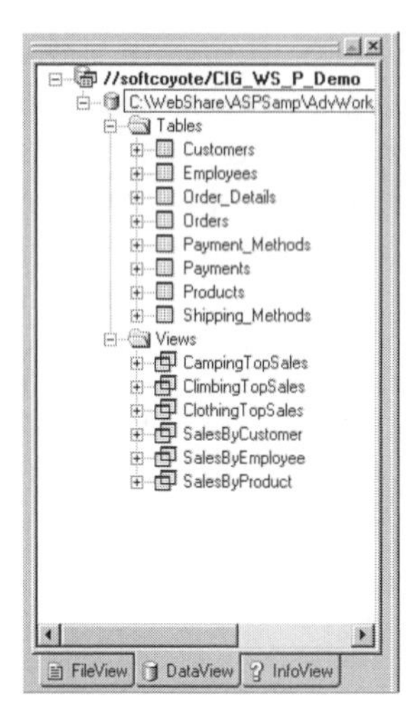

Creating a Project

With Visual InterDev open, choose **File**, **New** and click the **Projects** tab of the New dialog that appears. Highlight the **Web Project Wizard** option and enter a name for the project in the text box Project **Name**. (The figure below shows the name "CIG_Chap6_01" for the project name.) The default location for the project files is shown in the Location text box. (It is C:\Program Files\DevSTudio\MyProjects.) You can change this, but I prefer to leave it as is—this is because all of my projects are in one place in separate directories.

Since we don't have a workspace open, there is no choice except to leave the radio button selection set to **Create new workspace**. Now click the **OK** button and the Web Project Wizard takes over and walks us through the creation process.

Make sure that everything is correct before you click the OK button, since so many things are created and changed at this point.

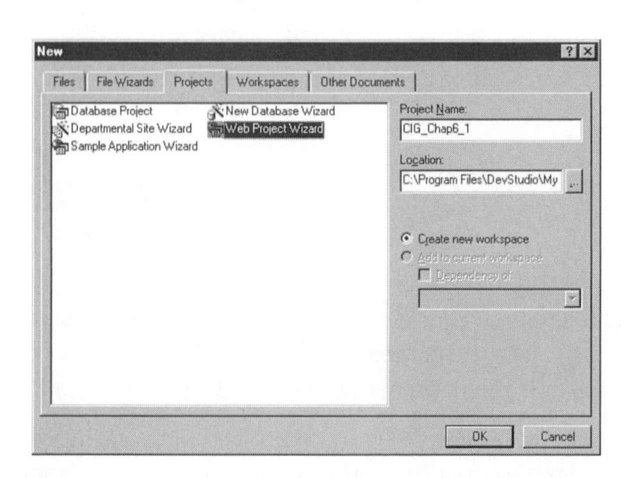

The Web Project Wizard now moves to step 1 of 2—naming the Web server upon which the project will be created. (Note that there is an option to use SSL, or Secure Sockets Layer, for the Web that is being created.) The Web server can be on the same machine from which you are running Microsoft Visual InterDev or it can be on another machine. (I have done both and it is difficult to see any difference.) Enter the name of the Web server as shown in the following figure and click **Next**.

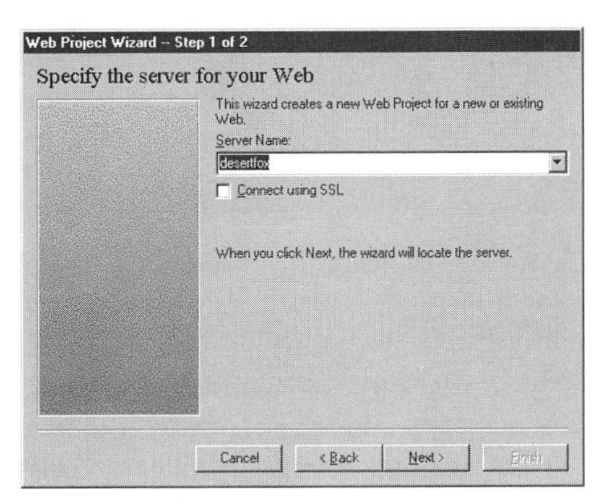

If you realize that you have done something wrong, you can click Back or Cancel.

Secure Sockets Layer

Because transmission of data on the Internet is open and subject to being examined by anyone with spy software and a desire to snoop, there has been a reluctance to send any sensitive information over the Internet without encryption. Encryption usually requires that the sender and receiver both have the key to the code. Secure Sockets Layer is a facility that is available in Microsoft Internet Information Server that simplifies the process of encryption and decryption of data so that sensitive data can be sent over the Internet without both the sender and receiver having an encryption key.

Create a New Web?

When the Web Project Wizard moves to step 2 of 2, you are asked whether you are creating a new Web or if you are going to attach this project to an existing Web. The next figure shows the New Web being selected.

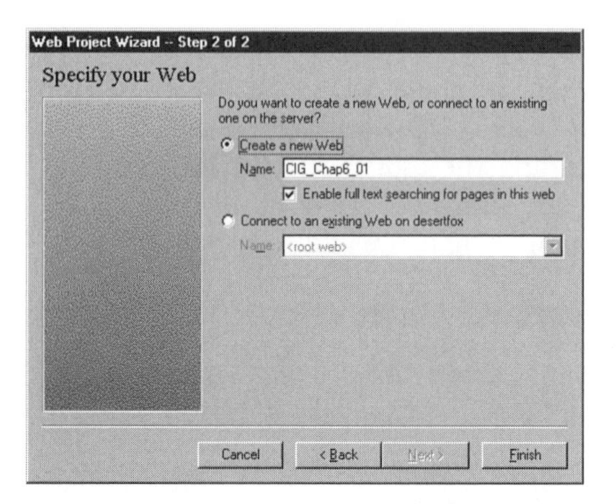

The Project name that was selected earlier is the suggested Web name.

Full Text Search

You are presented with the option of enabling a full text search capability for your Web. If you say yes to this option, a page is created named Search.htm. This page is powered by a FrontPage WebBot, which is part of the FrontPage extensions that were installed. The specific DLL that is used is SHTML.DLL, which you will find in the _vti_bin directory of a FrontPage-enabled Web. What the full text search enables you to do is enter a word or phrase in the query and have all of the pages of the Web searched for the word or phrase anywhere in the Web. One warning, if you have a large Web, searches can take a long time and a lot of server processing resource.

Now, click **Finish**. The Web Project Wizard finishes creating the project and basic Web structure, and adjusts directory settings on the server (as required). When it finishes, you will see the FileView tab has been added to the workspace window. When you click the plus symbol (+) to the left of the entry **//servername/CIG_Chap6_01**, you will see the files that have been created.

You can now edit any file in the project. If you double-click a file name, the file is opened in a text editing window as shown in the following figure.

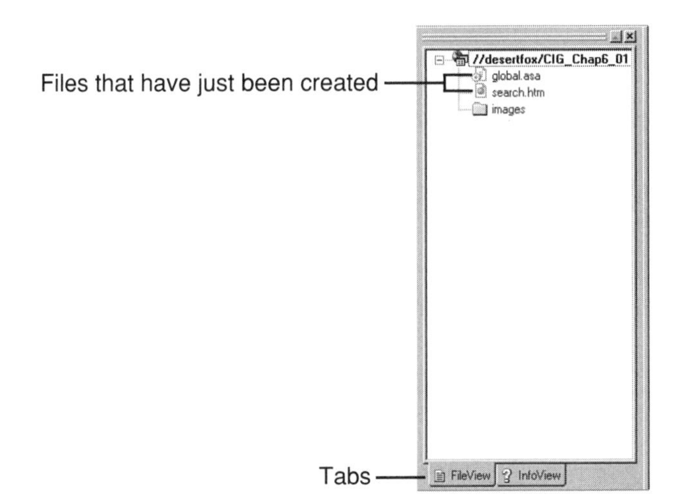

If you click the InfoView tab, the documentation is made available to you.

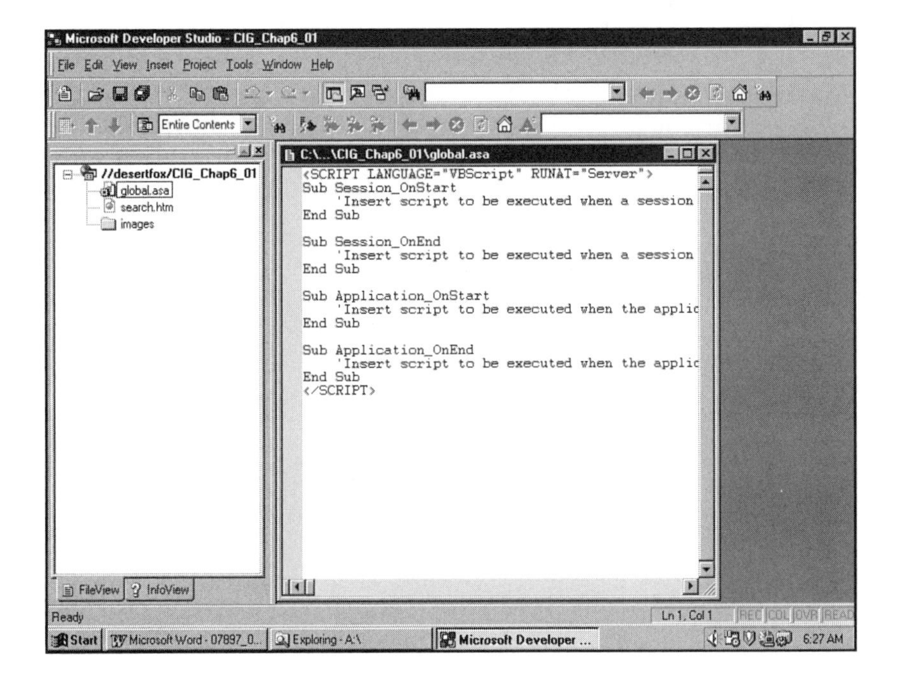

There is only one global.asa file for each Web site or application on the Web server.

Global.asa

The Active Server processes the global.asa file when you open or close an application or start or end a new Web session. You can include scripts that will run either at the start of a session or the opening of the application. An example of what you might want to do is to run scripts that open and close a database connection. There is a fair amount of overhead entailed in opening a database connection so you should keep it open until you are finished with it.

Edit With Visual InterDev Only

Microsoft Visual InterDev keeps copies of the project files in two places. When you change one of the files through Visual InterDev, both copies are changed. For this reason, it is extremely important to remember that you should always edit the project files through Visual InterDev. Opening one of the copies with a text editor such as Notepad and making a change can produce unpredictable results.

Working Directory and Working Copies

When the project is created, the default location for the files is in a folder named MyProjects in the following path:

```
c:\Program files\DevStudio\MyProjects\
```

A folder is created here with the name that we have assigned to the project, in this case CIG_Chap6_01. In the following figure, you see that a folder and five files are present in this folder.

Working Copies

The two copies of the files are kept for a reason. The copies in the MyProjects directory are the working copies of the file. These are the copies that you can work and experiment on without changing the Web site copy until you are ready to "release" the file to the Web.

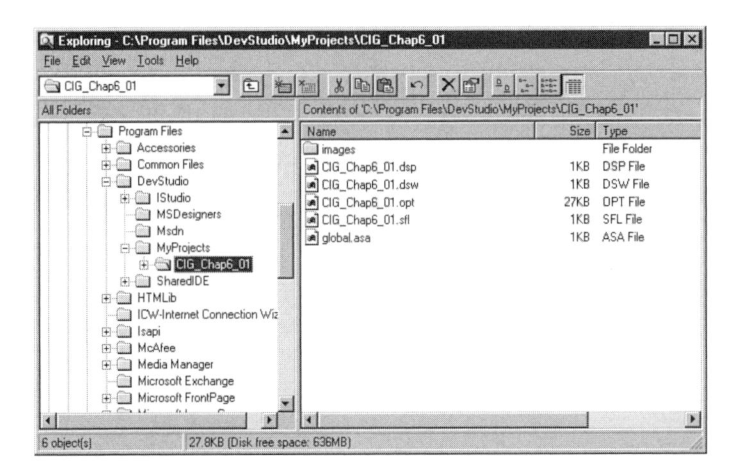

The images folder is created to contain any image files that you use in your Web project.

I know you are thinking that the workspace window showed two files and only one of them is here—the global.asa file. The global.asa file is here because we opened it earlier for editing. Prior to editing it, it existed only in the Web site.

The other file, Search.htm, would be here if we had opened it for editing. (It will be here when we open it for editing.) Two copies of this file exist—we will look at the other folder in a minute.

Right now, we need to understand what the other four files in the Project folder CIG_Chap6_01 are, and what we need to do with them. They are the files that Microsoft Visual InterDev uses to keep track of the project and workspace:

- The CIG_Chap6_01.dsp is the project file.
- The CIG_Chap6_01.dsw file is the workspace file.
- The other two files, CIG_Chap6_01.opt and CIG_Chap6_01.sfl, are binary files used by Visual InterDev. It is very important to understand what you should do with these files—ABSOLUTELY NOTHING! Leave them alone. Altering them in any way can make your project or workspace unusable. Trust me, I tried it for you. I had to dump a perfectly good project because I wanted to see what would happen if I modified one of these files.

Web Directory and Web Copies

Now it is time to look at the Web server folders to see where our project files are. The location of your Web server root file is probably one of two places depending on whether you are using the Personal Web Server for Windows 95, the MS IIS, or the Peer Web Server for Windows NT:

- ➤ If you are using Windows 95 Personal Web Server, you should find a directory named Webshare on the root directory of your C:\ drive. In this directory, there will be a directory named Wwwroot. This is the root directory of the Web.
- ➤ If you are using MS IIS or Peer Web Server for Windows NT.4, the directory will be InetPub located on the root of the C:\ drive if you installed in the default location. Within this folder will be a folder named Wwwroot.

Whichever Web server you are using, you will find the project folder CIG_Chap6_01 located in the Wwwroot folder. When you open this folder, you will see the project files as shown in the following figure.

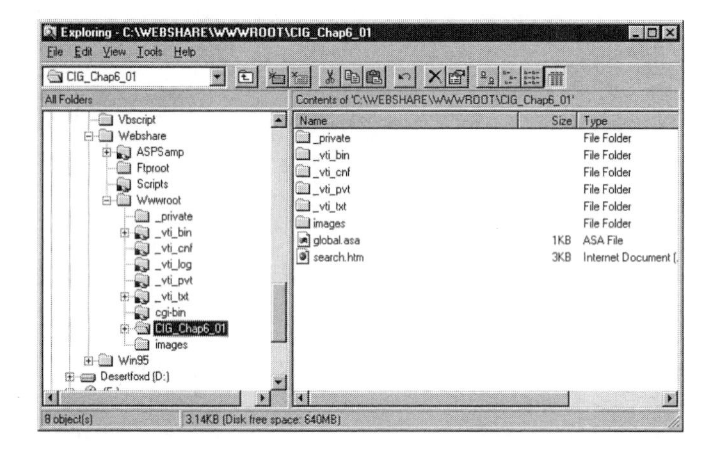

The standard set of folders (such as the images folder) are also created for the project.

Simply creating the project folder CIG_Chap6_01 in the Wwwroot directory is not enough for it to be recognized by the Web server. For the project to be recognized as a Web by the Web server, the directory needs to be set up in the Web server administration. Guess what? Visual InterDev did this for us when the project was created.

Visual InterDev has clearly performed much of the heavy lifting and attended to the little details involved in the creation of a Web site. You are left with the creative part of Web development, the fun part.

But before we open our new Web site for the first time, we need to create a file in our project.

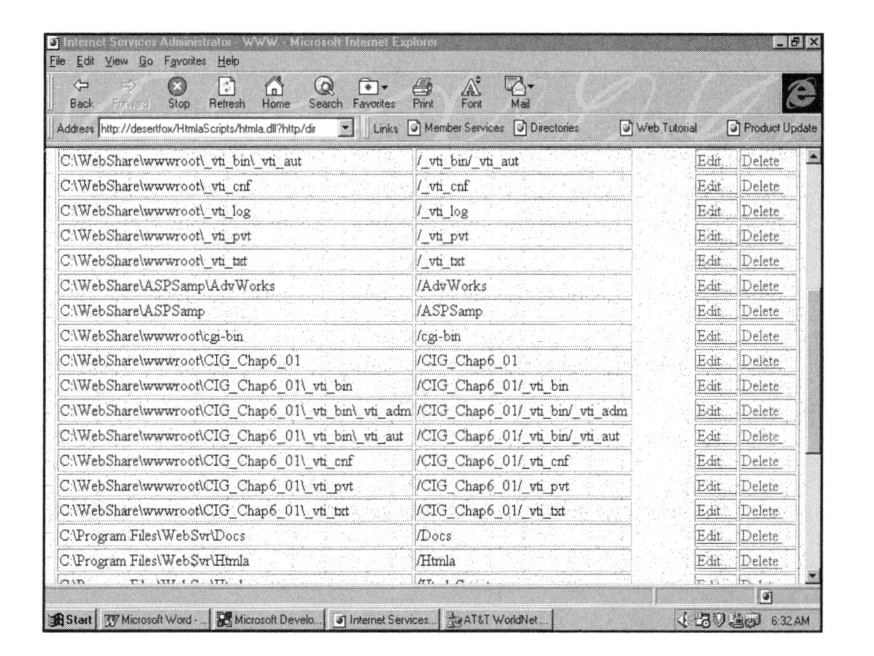

The folders are already set up—each with their own aliases.

Adding and Editing Project Files

Almost every Web site has a default page—this is the first Web page the server sends when a user visits your site. You can name the page anything you choose and that your Web server will support. The most typical names for this start page are Default.HTM and Index.HTM or Index.HTML. Default is used by default (do you think we have enough "defaults" running around?) by Microsoft web servers. Index is used by most other Web servers.

There are no hard and fast rules here. To "finish" our site, we are going to add one page named Default.HTM. We will then edit that page and add one link to the Search.HTM page in our site. Finally, we will pay a visit to our site with our Web browser.

Adding a Page to the Project

First, we need to open our project if it isn't already open. Start by opening Visual InterDev. Now choose **File**, **Open Workspace**. This will display a file dialog box that will open in the folder MyProjects, where you should see the folder CIG_Chap6_01. Double-click this folder icon and highlight the file **CIG_Chap6_01.DSW** and click **Open**. The project is opened where we left off before.

To add the page Default.HTM, choose **File**, **New** on the menu and click the **Files** tab. Select **HTML Page** and enter the name **Default** as shown in the following figure.

When the Add to Project check box is checked, the file is automatically inserted into the project.

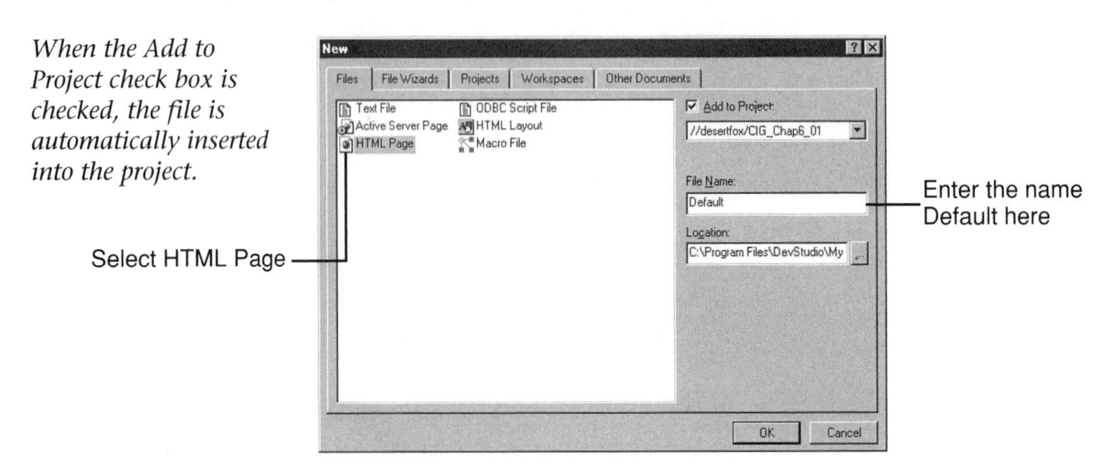

Click **OK** and the new file is created. The new file, Default.HTM is now displayed for editing with the line `<!-- Insert HTML here -->` highlighted and ready for our editing, as shown in the next figure.

The <! Notation starts a comments line that won't be displayed by the Web browser.

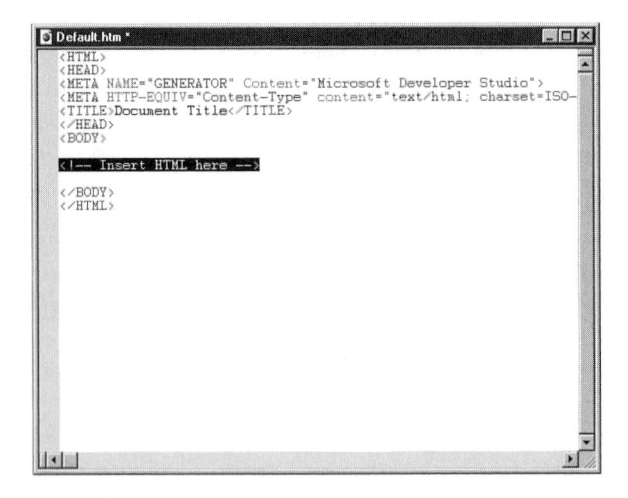

Editing a Project File

So let's edit the file by adding some text. Add the following in place of the highlighted line:

```
<H2>This page was created by the intrepid developer</H2><BR>
<H1>Your Name</H1><BR>
<H2>Who is now an experienced Web Developer</H2>
```

Now save the file by choosing **File**, **Save** in the menu. We can now preview our handy work as it will appear in a browser. To do this, you don't have to open your browser. Simply move your mouse pointer over the document edit window and right click. A floating menu appears. Click **Preview Default.HTM** and you will see a WYSIWYG view of your page in the InfoViewer Topic window, as shown in the following figure.

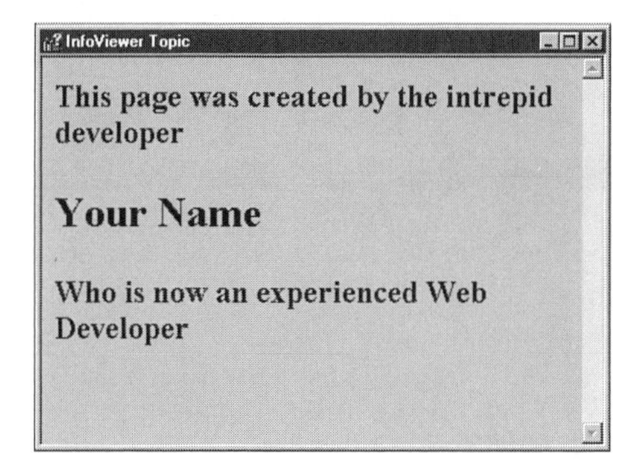

The ability to quickly preview a page as it will appear is a great convenience.

Now it is time to add the link to the Search.HTM page that was mentioned earlier. All that is required is to add this line of code to what is already there.

```
<BR><A HREF="Search.HTM">The Search Page</A>
```

Checking the Link

Now save the page, open your browser, and set the URL to **http://Servername/CIG_Chap6_01**. You will see your Default.HTM Web page displayed with "The Search Page" as a link to another Web page. When you click the link, you will see the Search page. Experiment with this before you go on to other tasks.

The link appears in a different color. This color is set in the browser. Most browsers change the link color after the link has been visited.

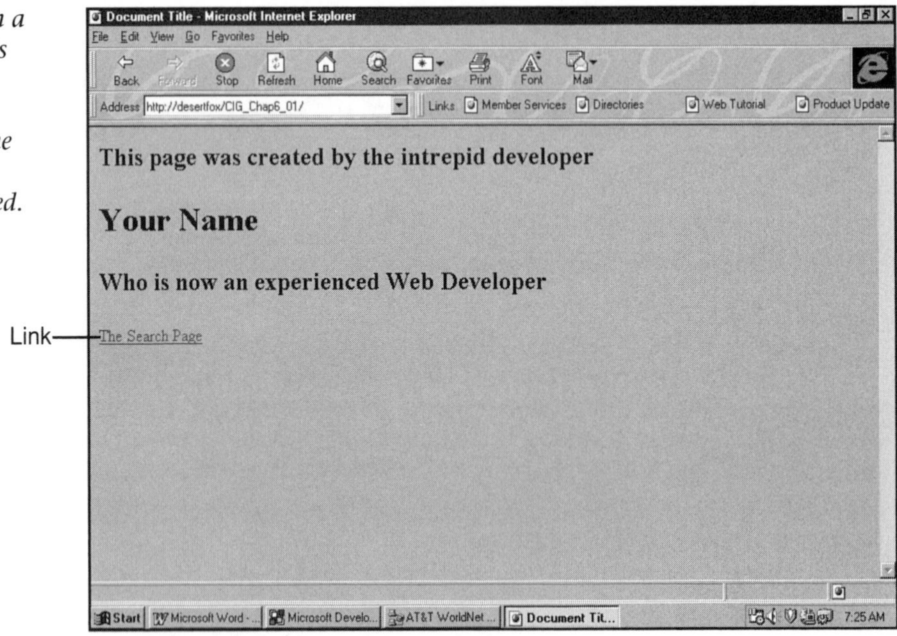

You have now created a simple but complete Web site. You have control of what is arguably the most powerful Web development tool in existence today. And you have experience.

Removing a Project

Normally you won't want to remove a project as soon as you have completed it. These instructions are here so that when you want to remove a project, you will be able to do so safely. Removing a Web project is not as simple as just deleting files. To properly remove a project and eliminate any unforeseen consequences, there are four steps.

1. Use the Web server administration function to remove the directories from the Web site. This is not the same as physically deleting the files. That comes later. Refer back to the figure that shows the Web server directories.

2. Stop the Web server before deleting the files from the Wwwroot directory.

3. Use the Windows Explorer to delete the files in the directory \Wwwroot\CIG_Chap6_01. Delete the directory CIG_Chap6_01. Do not delete the directory Wwwroot or any other directory. You should now restart the Web server.

4. Use Windows Explorer to delete the Directory C:\Program Files\DevStudio\MyProjects\CIG_Chap6_01 and all of its contents.

You have now safely removed the project and workspace.

Creating the Sample Projects

There are two sample applications that come with the Visual InterDev installation CD. These provide very good sample applications from which you can examine code, and are there for you to rip apart to see how someone else accomplished a task. These are marvelous reference sources—use them.

In this section, we are only going to install these applications. You will have the fun and adventure of exploring them yourself.

Installing Dos Perros (Two Dogs) Chili Company

Open Visual InterDev and choose **File**, **New**. Click the **Projects** tab and highlight the **Sample Application Wizard**. Enter **Dosperros** for the project name as shown in the following figure.

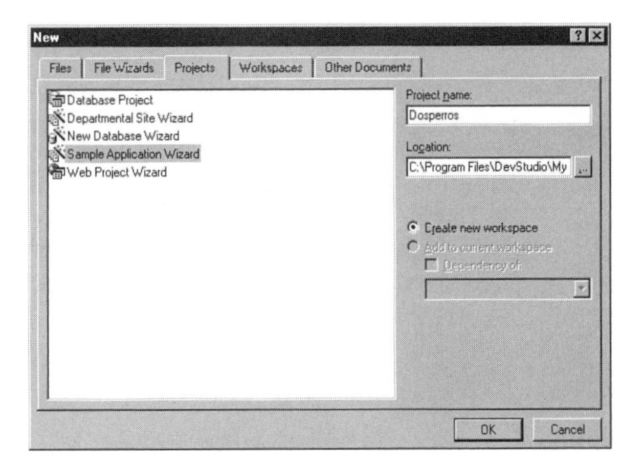

The choice of Create new workspace is the only choice since a workspace is not open.

Click **OK**. This opens step 1 of 5 of the Sample Application Wizard. Select "DosPerros Tutorial Application" and click **Next**.

You will be asked various questions by the Sample Application Wizard as you proceed through the steps. Note that in steps 3 and 4, you will need to create a directory and make the directory a share with full rights (since the Dos Perros sample includes a database; for those applications with no database, steps 3 and 4 are skipped).

This Wizard can also be used to install a custom application that is supplied with a INF file.

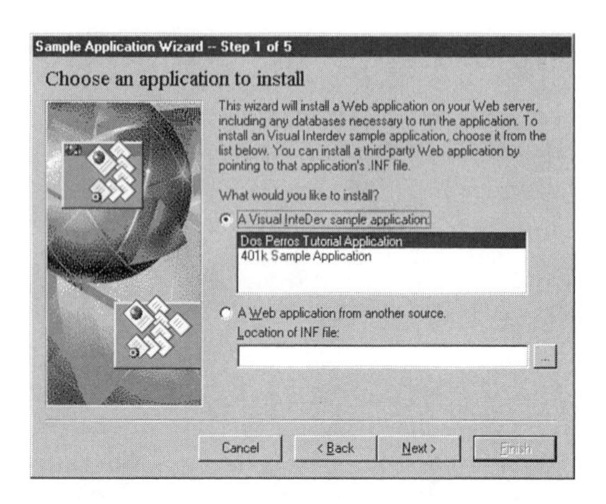

Microsoft Access Not Required

You will notice that this sample application uses an Access Database. It is not necessary that you have Microsoft Access installed on your system, just that you have the Access ODBC driver (which was installed when you installed Visual InterDev).

Step 4 of 5 is where you will supply the directory name that you created before we started this installation. As you see in the figure below, it is entered in the UNC format of \\Server\Sharename rather than the more familiar C:\directory format.

After step 5, the Sample Application Wizard will report, "That's all we need!" At this point, click **Finish** to install the sample application. When the workspace is opened, notice that there is a Default.HTM file. The URL for opening this project is **http://Servername/DosPerros**.

If the share has not been properly created, you will receive an error message. Simply create or correct the share and continue.

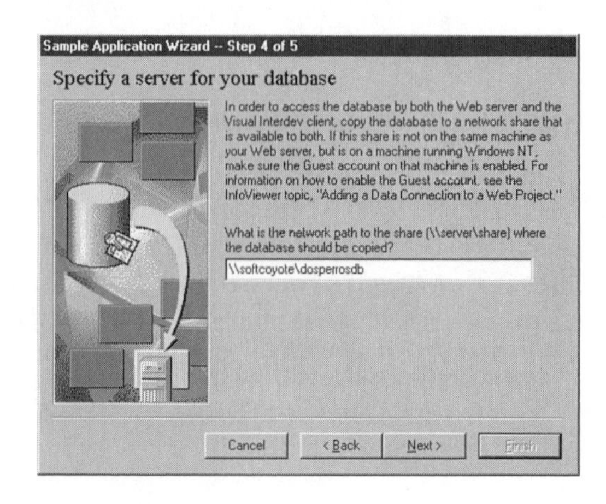

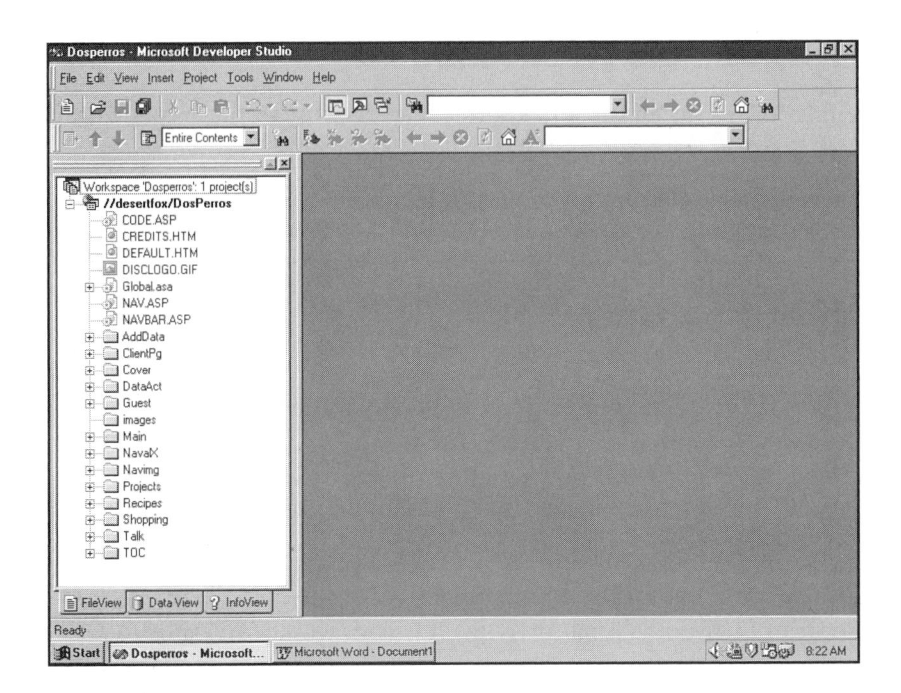

Notice that there are also some ASP (Active Server Pages) here.

Since there is a database, a DataView tab has also been added to the workspace. Open the directories and you will see the tables as shown in the following figure. (Databases are discussed in detail in Part 4, "Databases.")

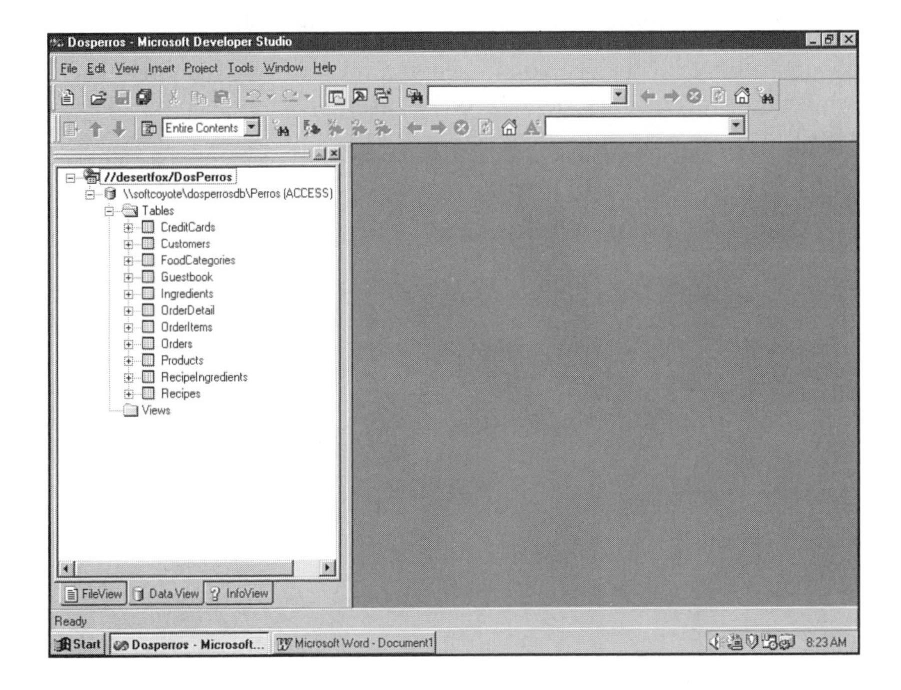

If you double-click the table name, the table is opened in a grid window so that you can see the contents.

Installing the 401(k) Sample Application

The steps that you will follow to install the 401(k) project are like the steps for the Dos Perros project. The URL is **http://servername/401k**.

I recommend that you seriously examine this project. It combines the use of VBScript on the server, JavaScript for the client, and the use of ActiveX controls.

The Least You Need to Know

In this chapter you have explored the functional heart of Microsoft Visual InterDev. Despite the many features that were examined, there are many more to be mastered. The best way to learn these tools in depth is to use them. The tool is only as effective as the person using it. (And the person using it is only as effective as the experience and effort expended mastering the tool.) If I haven't made my point yet, Practice, Practice Practice! In the next chapter you will be creating your own project.

- The workspace is the container for the entire project. All of the files can be edited from the workspace of Microsoft Visual InterDev. It is your workbench and workshop.
- The MyProjects folder contains the working copy of the project and files. These are the copies that you "Work On."
- A workspace can contain more than one project. This lets you divide a Web into pieces for more convenient handling during creation and maintenance.
- The Microsoft Visual InterDev will assist you in keeping track of the elements of a very complex project. One of the great benefits of Visual InterDev is keeping control of all of the details.

Part 3
Active Server Pages

Active Server is one of the very exciting developments to come from Microsoft in the World Wide Web development arena. Prior to Active Server, developers had very limited choices if they wanted to create dynamic Web pages.

The first set of choices depended upon the features supported by the client Web browser.

The second set of choices were the use of CGI (Common Gateway Interface).

Then came Active Server (drum roll please). All of the processing is performed on the Web server. This means that the Web developer and publisher is not dependent upon the features of the browser used on the client system. Only standard HTML is transferred to the client system. Active server supports Active Data Objects, ActiveX, VBScript, JScript, and all future scripting languages. So pull up a chair and watch. This is fun stuff.

TRAPPED IN A BROKEN ELEVATOR WITH A TECHIE...

Chapter 7

Active Server Pages—Don't Just Sit There!

In This Chapter

- Understand why everyone is so excited about Microsoft Active Server
- Learn to explain the difference between ActiveX and Active Server (Warning: Repeating this explanation may result in glazed eyes in your audience.)
- Understand the significance of client browser independence
- Use Visual InterDev to create a simple Active Server page
- Understand how to use scripting (both VBScript and JScript) with Active Server
- Understand that a file extension of .ASP refers to an Active Server Page, not Cleopatra's pet snake

Check out the CD-ROM!

This chapter has corresponding sample files on the included CD-ROM. Simply click on "Examples," then on the corresponding chapter number in which you are interested (be sure to read the "Readme" text file, also located in the section "Examples", for important information on installing the sample files).

In the beginning, Microsoft created VBScript. Microsoft hailed VBScript as terrific (and so did many others) because it allowed interactive content in a Web page using Visual Basic-like script commands. This made it possible for anyone who knew Visual Basic to use VBScript with relative ease.

Microsoft then created ActiveX. ActiveX makes available many previously-created ActiveX controls that need only be plugged into your Web page. Easy for the non-Geek. (Prior to this, interactive Web content required knowing Java, PERL, C++ or some other more difficult-to-learn-and-use technology. This meant that only geeks had truly exciting Web pages).

ActiveX New or Recycled

ActiveX Controls are a revision of OLE Controls. While Microsoft hails them as new in one breath, they are explained as having a long and distinguished parentage. So I guess that they are both new and old—sort of recycled. For more about ActiveX Controls, refer to Chapter 10, "ActiveX Controls, OLE & VB5CCE: Making your Own."

However, there is one small problem with the use of VBScript and ActiveX in your Web page. The Web browser being used by the Web page visitor is required to support VBScript and ActiveX. When a Web browser supports ActiveX or VBScript, it is able to understand and act on the VBScript commands and recognize the ActiveX control. When a Web browser doesn't support ActiveX and VBScript, it ignores it. Since the only current browser that supports both ActiveX and VBScript is Microsoft's Internet Explorer, a Web site visitor using the Netscape browser will see none of your work with VBScript and ActiveX controls.

This is where Active Server comes to the rescue and solves the problem.

Active Server—the Solution to Web Evolution

Active Server supports ActiveX controls and VBScript, but it provides this support on the server; in other words, it doesn't depend on the Web browser for support. Since Active Server sends an HTML page directly to the browser, it doesn't matter which browser is being used—all your cool ActiveX and VBScript will still work!

In the current realm of Internet and particularly World Wide Web development, one of the problems faced by developers is rapidly changing technology and standards. Fueled by the growing expectations of the Web visitor, the pressure for high quality, interactive, state of the art Web pages is enormous. The creation of Active Server is a response to this world of ever changing standards.

Understanding Microsoft Active Server

Microsoft Internet Information Server 3.0 does not replace or change any of the functionality of Microsoft Internet Information Server 2.0. What MS IIS 3.0 *does* accomplish, however, is to add Active Server functionality to MS IIS 2.0. The Active Server can also be used with the MS Peer Web Server (which is available with MS Windows NT Workstation) and the MS Personal Web Server for Windows 95. It is the same installation set of software in all cases and the functionality is identical.

Active Server Intallation

The installation of Active Server pages requires a few things. First, MS IIS 2.0, MS Peer Web Server or Personal Web Server for Windows 95 must be installed.

Also, the installation of Microsoft Internet Explorer 3.01 for Windows 95 or Windows NT 4.0. is necessary. The Build required is 1215 or higher. You can check the build of your version of Internet Explorer by opening the Internet Explorer and choosing **Help**, **About**. MS IE 3.01 is usually supplied with the MS IIS 3.0 install set. In the event that it isn't supplied, it can be downloaded from the Microsoft Web site at **http://www.microsoft.com.**

When MS Active Server is installed, an icon titled Active Server Pages Roadmap is added to the MS IIS, Peer Web Server or Personal Web Server program Group.

When you click the Active Server Pages Roadmap icon, documentation and samples of the use of Active Server are opened in your Internet Explorer browser as shown in the following figure.

The TCP/IP Stack

TCP/IP Stack must be running for the Active Server functions to work. An easy method of starting the stack is to attach to your Internet Service Provider just as if you were going to browse Web pages. If your Internet Explorer is set to automatically connect with your Internet Service Provider when you open it, the same will happen when you click the Active Server Pages Roadmap icon. Simply allow the connection to be made and the stack will run.

If you receive an error informing you that the pages or site can't be found or that there is no program associated with the .asp file extension, the probable cause is that the stack is not running.

The frame on the left contains links to the documentation on Active Server. When a topic is chosen, it is displayed in the frame on the right.

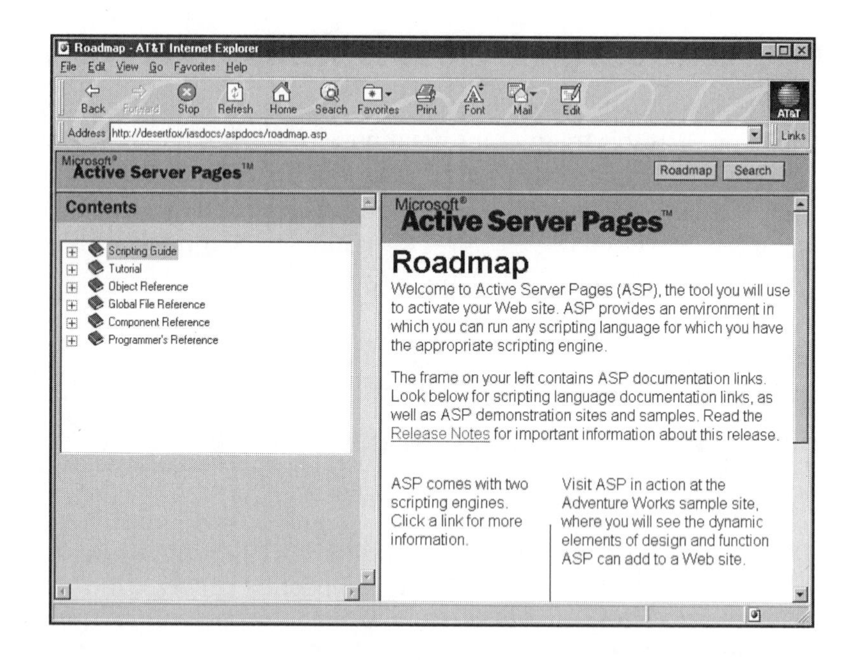

How Does the Server Know To Be Active?

When a client browser sends a request for a page to a server, the server sends the page. The server doesn't make any decisions about how to format the page, nor does it adjust the format of the page transmitted based upon the Web browswer making the request—it simply transmits the page.

What is needed is a signal to the server that you want the server to do something other than simply transmit the page. When the page requested has an .ASP file extension, the Active Server reads the page, acts on the instructions, and creates a Web page of all HTML that is sent to the browser. The Web page transmitted by the Active Server does not contain any ActiveX controls or VBScript; it only contains static HTML, which can be understood by almost any Web browser.

Creating an Active Server Page

The best way to understand an Active Server Page is to create one and see what it does. Begin by opening Visual InterDev (If the Tip of the Day appears, click the **Close** button). You should be looking at something very much like the next figure.

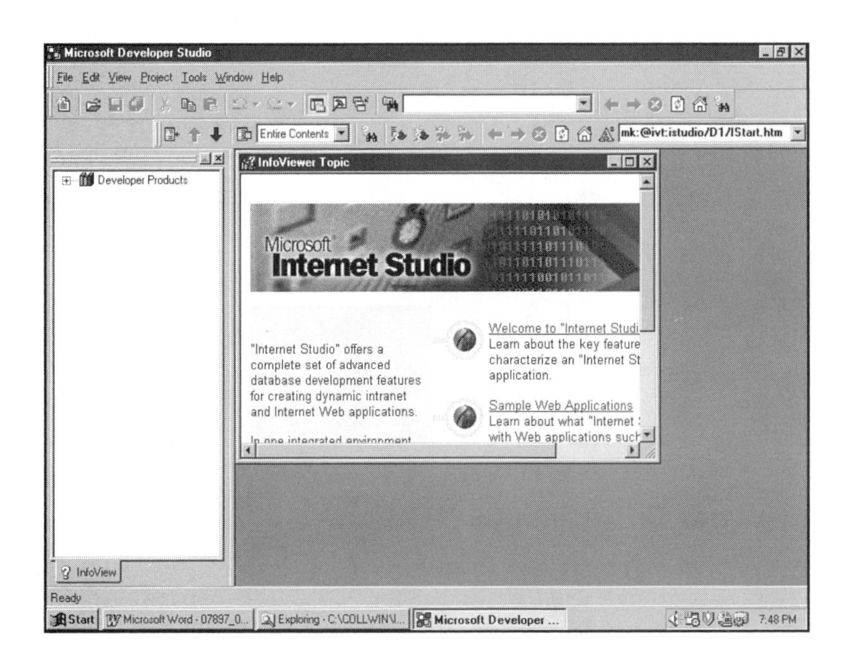

You are now ready to create a project.

Now that Visual InterDev is running, your next step is to create a project. Choose **File** on the Menu Bar, then choose **New** and click the **Project** tab. Enter the name of the Project that you are creating. (In the next figure , the name `CIG_Demo` has been entered). Set the radio button to **Create new workspace**. Note that when you do this, Web Project Wizard is selected.

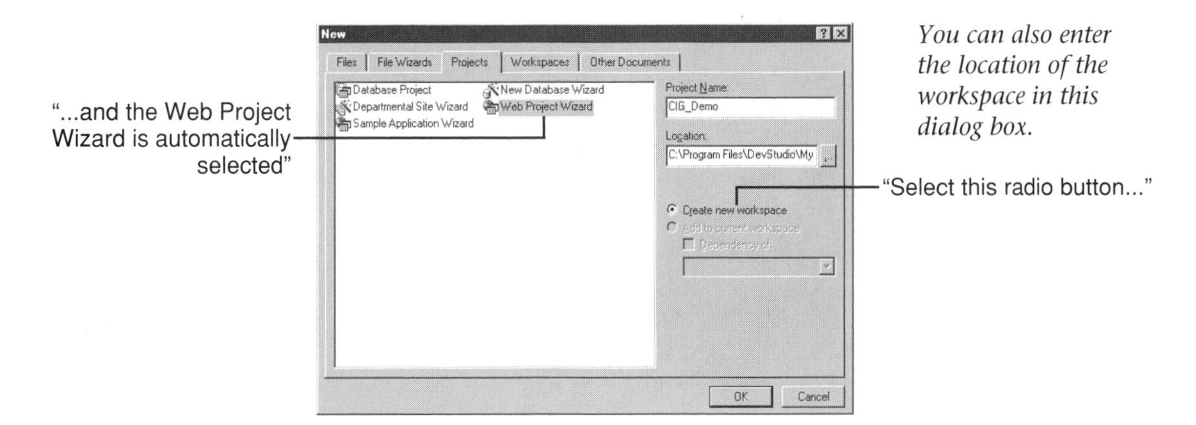

You can also enter the location of the workspace in this dialog box.

When you click the **OK** button, the CIG_Demo project is created by the Wizard. You are asked for the server name as in the following figure. Enter the server name and click the **Next** button.

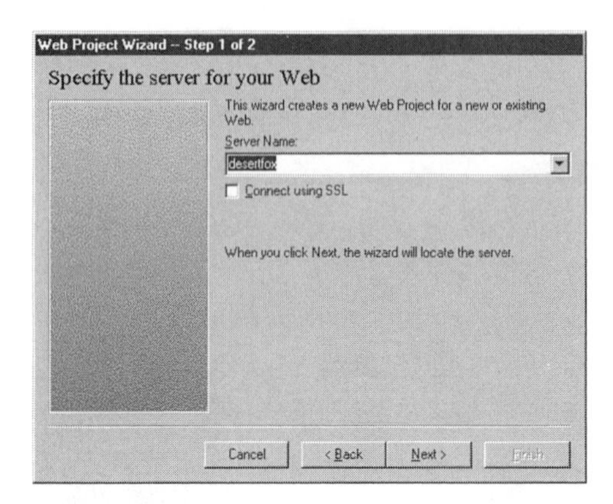

If the server name isn't shown, enter it in the textbox.

Finding Your Server (Computer) Name

If you don't know the name for your server and you are using the Web server on your local computer, you can check the name by choosing Start, Settings, Control Panel. Click the **Network** icon and click the **Identification** tab. The Computer Name will be the Server Name.

In step two of the Web Project Wizard, leave the setting to **Create a new Web** as shown in the next figure and click the **Finish** button.

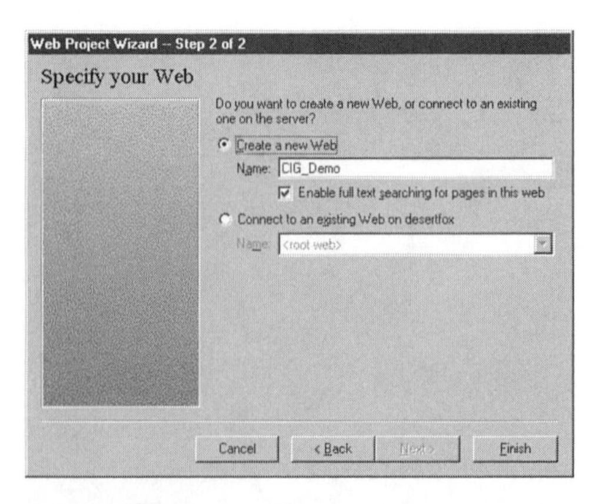

The project will now be finished.

Two files were created by the Project Wizard, Global.ASA and Search.HTM, neither of which we will need at present. This project was created in a folder named C:\Program Files\DevStudio\MyProjects\CIG\Demo. In the process of creating the project, a virtual directory /CIG_Demo was created on the Web server with execute permission.

Web Server Virtual Directories

A Web server virtual directory has two locations. The physical location on the disk is the type of path with which you are most familiar. It might be C:\directory\file.htm for example. The *Web virtual directory* is defined in the Web server and represents the logical relationship of the directory relative to the Web root directory. For more information, refer to the documentation for the Web server that you are using.

The Global ASA

The function of the Global ASA file is to provide session or application procedures for starting or closing. An example might be opening a connection to a database and closing a connection to a database each time a session is started. There can be only one Global ASA file per application. It is always named Global.ASA. It must reside in the root of the application directory. Since you can do everything in the .ASP that you do in a Global ASA, these files are not required but are most useful for a large application. They can eliminate redundancy and remove a source of error by having code on only one location.

Creating the Active Server Page

It is now time to create the Active Server page. From the main Visual InterDev menu, choose **File**, **New**. Click the **Files** tab and select the **Active Server Page**.

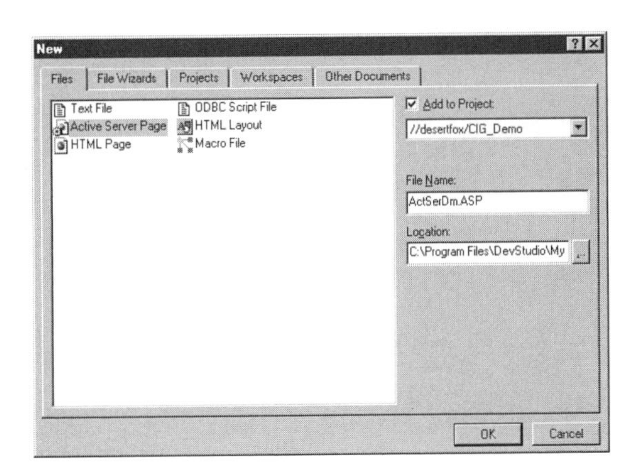

You are now ready to create an Active Server Page.

Make sure that the **Add to Project** box is checked and enter the name **ActSerDm.ASP** in the **File Name** text box. Click **OK**. Visual InterDev then creates a skeleton page that is ready for the VBScript, as you can see in the following figure.

Visual InterDev creates the boiler plate entries in the Active Server Page.

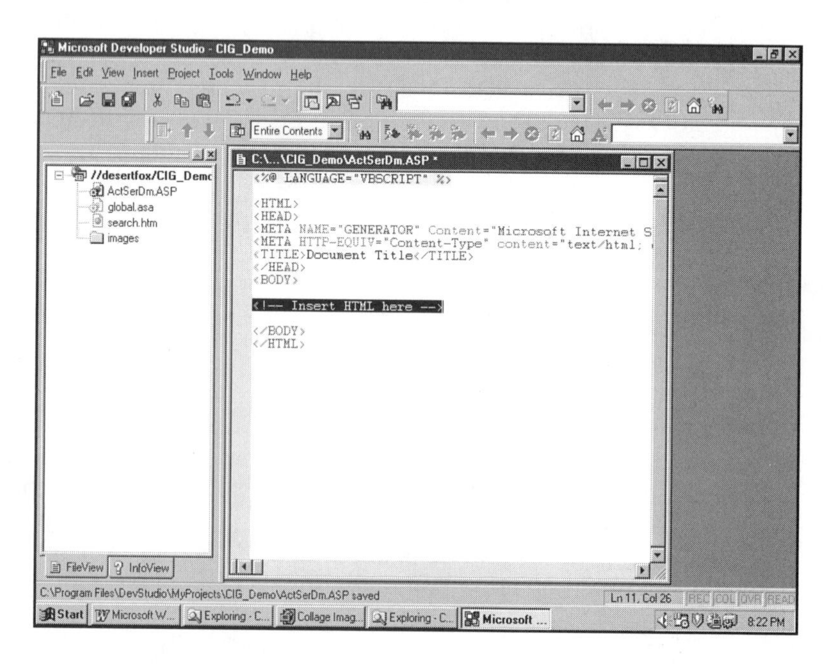

Program Logic for the Active Server Page

Technicolor Hints When your Active server page is open in Visual InterDev, notice that the text in the file is color coded, depending on what type of code it is. As an example, the VBScript is black characters on a yellow background. It is a great boost to readability.

You are now ready to add program logic. The following remarks line is highlighted:

```
<!-- Insert HTML here -->
```

You will enter your VBScript and HTML in place of this line. Enter the following code exactly as shown in place of the remarks line:

```
<% for i = 3 to 7 %>
<font size=<%=i%>>This is a test of an Active
Server Page</font><BR>
<% next %>
```

Your Active Server page should now be ready for test.

What the Script Does

Before you look at the Active Server Page in action, you may want to know what the VBScript that was entered will do. Examining it line by line, the first part is the following line:

```
<% for I = 3 to 7 %>
```

As you know, all tags are enclosed in < > symbols. Next, you see that all VBScript instructions are enclosed in % % symbols. (The stuff in the middle does the work). The instruction creates what is known as a FOR loop. The variable I is created and given an initial value of 3. The value of I will be increased by 1 each time the loop is processed. The loop will keep executing until the value of I reaches 7.

The second line is:

```
<font size=<%=i%>>This is a test of an Active Server Page</font><BR>
```

The first part sets the HTML Font size equal to the variable I. The <BR> creates a line feed.

The third line is:

```
<% next %>
```

This line causes the loop to be repeated.

If your Active Server page works properly, the page will show the line, `This is a test of an Active Server Page`, five times on five separate lines, each larger than the last.

Check This Out...

VBScript in Depth For more on this topic see *SE Using VBScript* by Ron Schwarz and Ibrahim Malluf.

Testing the Active Server Page

Testing the Active Server Page that you have created will help you understand what happens with Active Server Pages. During the course of the test, you will see three separate versions of the Active Server Page.

Active Server Page (ASP) Version

The first version that you will see is the contents of the ASP file. To see this version, you can open the file ActSerDm.ASP with Notepad. The following figure shows the ASP version of the page.

Editors

In addition to the text editor in Visual InterDev and Notepad, the FrontPage Editor that comes with Visual InterDev is available for your use. Refer to Chapter 17, "Working With FrontPage 97: What You See."

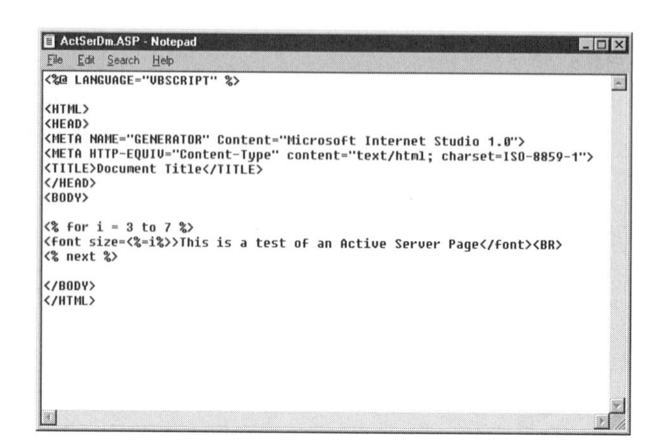

```
<%@ LANGUAGE="VBSCRIPT" %>

<HTML>
<HEAD>
<META NAME="GENERATOR" Content="Microsoft Internet Studio 1.0">
<META HTTP-EQUIV="Content-Type" content="text/html; charset=ISO-8859-1">
<TITLE>Document Title</TITLE>
</HEAD>
<BODY>

<% for i = 3 to 7 %>
<font size=<%=i%>>This is a test of an Active Server Page</font><BR>
<% next %>

</BODY>
</HTML>
```

You can use to Notepad edit all Web page source text files.

The Browser Version

In order to view the Active Server Page with your Web browser, you will need to open your Web browser and set the URL to **http://servername/CIG_Demo/ActSerDm.ASP**. You should see a Web page somewhat like the following figure.

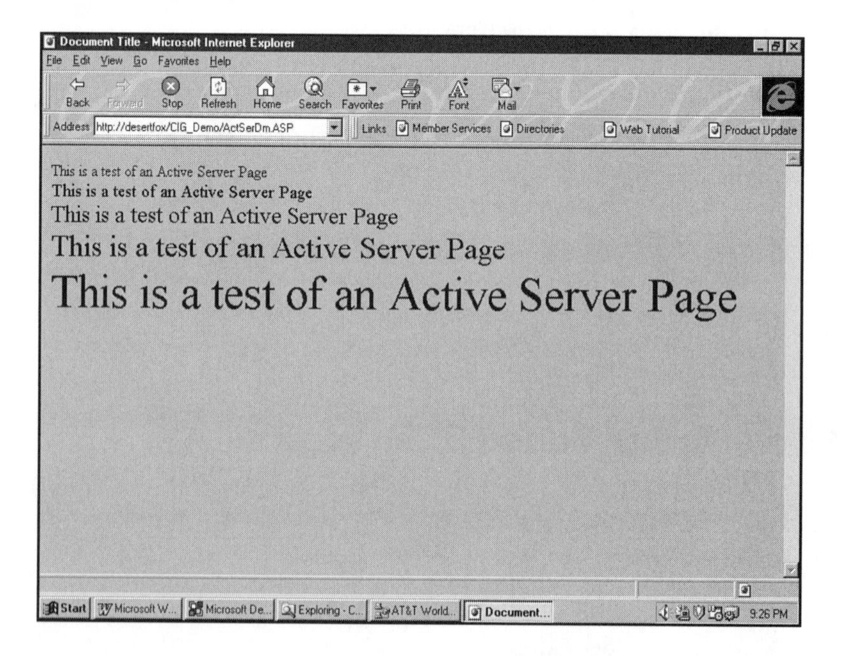

The increasing size of the font is a result of the value in the variable I.

The HTML Version

To see the HTML version, choose **View** in the Menu Bar of the Web browser and click **Source**. The source HTML file will be open in Notepad. As you see in the figure below, the HTML actually transmitted to the Web browser by the server is different from the Active Server Page version of the file. Read this paragraph again! It is important!

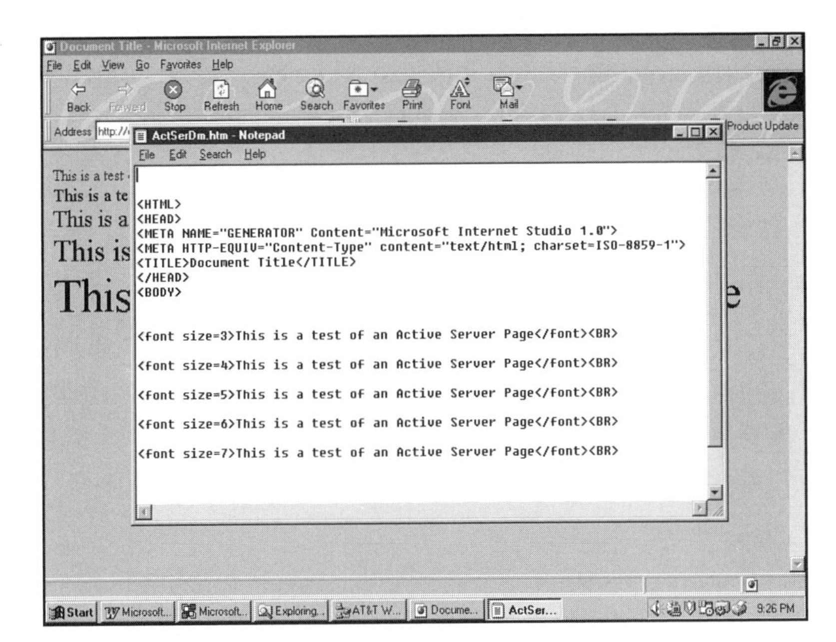

This is standard HTML that can be displayed by any Web browser.

The Active Server created the version that was received by the Web browser from the instructions in the Active Server Page. When you examine the HTML in the page that was transmitted, you will see all elements that are part of standard HTML. This page can be viewed using Netscape or any of the other popular Web browsers.

The Least You Need to Know

With Visual InterDev, you don't have to be an expert to create Active Server Pages. There are other aspects of Active Server that will be explored elsewhere in the book. These include such powerful elements as ActiveX and database. All that is required to use Active Server is the basic knowledge of how to create a page in the Visual InterDev and a little VBScript. You will be creating interactive Active Server Pages that will be the envy of every kid on the block. Just remember:

- With Active Server, you are no longer restricted by Web browsers. You can now put interactive Web content on your Web pages.
- VBScript is easy to learn and use. You don't have to have a plastic pocket protector to use VBScript. (If you prefer, you can use JScript).
- Active Web content is far more powerful and appealing than static HTML. You can create real applications that you control with Active Server.
- Static HTML is rapidly becoming a thing of the past. The trend will move away from client-side processing.

Chapter 8

VBScript: Lights, Camera, Action!

In This Chapter

- Learn the syntax of VBScript
- Convert VBScript data subtypes
- Create a Web page with VBScript functions
- Understand the use of HTML forms with VBScript
- Control the program flow of a VBScript program

Check out the CD-ROM!

This chapter has corresponding sample files on the included CD-ROM. Simply click on "Examples," then the number corresponding to the chapter in which you are interested (be sure to read the Readme text file, also located in the "Example" section, for important information on installing the sample files).

In this chapter, you will again be visiting the wonderful world of programming. So break out the plastic pocket protector, put on your propeller beanie, and set a techno-geek expression on your face. Most of all, don't be afraid of programming. Most programmers that I know are mere mortals (just don't tell them).

A member of the Visual Basic family of languages, VBScript shares the characteristic of all scripting languages in that it is interpreted. Some other important characteristics include the following:

VBScript Capability At present, Microsoft Internet Explorer is the only Web browser that has an extension for the interpretation of VBScript. Microsoft Active Server (also called MS IIS 3.0) is the only other VBSCript interpreter capable program.

- Much like HTML, VBScript cannot stand alone. It is always embedded in an HTML file and is comprised of ASCII text statements. ASCII text can be read by mere mortals. It never exists as an object code binary object (which can't be read by mere mortals) since it is not compiled.
- VBScript must be processed by an interpreter, which is either an extension to a Web browser for the interpretation of VBScript or an extension to Active Server for the same purpose. An interpreter is a program that reads the script and executes the instructions.

Why Use a Scripting Language?

I know, you were just getting the hang of HTML and here I am, stacking another language on top. There is a really good reason for introducing scripting languages into Web pages—to overcome the inherent limitations of HTML. Essentially, HTML consists of formatting commands and links to inline objects such as pictures and hyperlinks to other Web pages. HTML does not perform conditional logic. An example of conditional logic would be for the Web page to look at the computer's internal clock and if it is morning, send a greating of "Good Morning" at the top of the Web page. However, if it is afternoon, the Web Page would read "Good Day" (or "G'day Mate" if you are in Australia). HTML is not capable of manipulating variables such as adding two numbers together. In other words, it does not possess the functionality expected of a full programming language.

Compilation versus Interpretation Scripting languages are sometimes criticized for being slow in execution and for not being "real" programming languages because they are interpreted and not compiled. Actually, all compilation consists of is the interpretation of the program instructions into machine instructions. The interpreter does the same thing, it just does it at the time the program is being run.

Scripting languages have an advantage over compiled languages (especially in applications such as Web pages) in that the script is processed by another program that is responsible for the interface with the system resources.

Changing a VBScript statement in a Web page is as simple as editing any text file. No compilation or link-editing is required.

Other Scripting Languages

VBScript is not the only scripting language in use in the World Wide Web today. JavaScript and JScript, which is Microsoft's implementation of JavaScript, are becoming widely used. Microsoft is creating a standard interface to other scripting languages such as PERL and REXX. It is only a matter of time until other scripting languages will be introduced to the Web. Refer to Chapter 9, "JavaScript, Java, Java Applets: More Than Just a Cup of Coffee." You may also want to check out *The Complete Idiot's Guide to JavaScript, Second Edition*, by Aaron Weiss.

VBScript Language Elements

If you are familiar with Visual Basic in any of its versions, you will be right at home with VBScript. VBScript has all of the usual features and functions of a programming language to be considered a true programming language.

Since there is never a correct order to cover the elements of a language, let's just jump in.

VBScript Data Types

Data types distinguish between things like numbers and strings of characters that make up words. Integers (whole numbers) are distinguished from floating point numbers (numbers with a decimal point and a varying number of places after the decimal). There is also a distinction for currency (money) numbers. Currency always has a fixed number of decimal places and must contain precise numbers. Dates and times are also a different data type, as are objects.

It is true that VBScript has only one data type, which is called Variant. Variant, as its name implies, can contain various types of data. OK, then why are we talking about data types? Because VBScript, in an attempt to be sneaky and underhanded, uses subtypes. The type of data that is stored in a variable is recorded in the variable as a subtype.

Why this confusion? VBScript is what is called a loosely-typed language. When you create a variable, you don't have to decide what type of data it will contain. The subtype is determined by the context and the data that is actually placed in the variable. For example, if you have the number 100 stored in a variable of the Variant data type, when your code context treats it as a string, it behaves as a string. When you use it in a calculation it will behave as a number. VBScript will convert the data as necessary, depending on the context.

Refer to Table A on the Tear-Out card for a list of the Variant data type subtypes.

If you attempt to perform an operation on a variable (such as adding two variables together) you will get an error if one of them is a string subtype. This is true even of the self-converting Variant data type. It does not always work as expected. You will still encounter the occasional type-mismatch error condition. There are some VBScript functions that allow you to examine the Variant subtype and to convert the subtype to eliminate the type mismatch problem. As stated earlier, one of the consequences of the Variant data type is that when a variable is declared, the type is not specified.

VBScript Functions

VBScript contains the most of the functions that you are familiar with from Visual Basic. We aren't going to cover the full list of functions here since there are a whole bunch. There are some functions specifically for working with the Variant data type subtypes that deserve some mention. The first of these are two that are used to determine the subtyp- of a variable.

Several functions test for specific Variant subtypes. See Table C on the Tear-Out card.

There are functions that will convert the Variant subtype that will leave you with a known subtype. See Table D on the Tear-Out card.

The functions that you will not find in VBScript are Val() and Str(). You will need to use the CStr() or other Variant subtype conversion functions to accomplish the same result.

VBScript Variables

Declaring a variable in VBScript is very similar to Visual Basic. In Visual Basic, if you don't declare the type, the variable is a Variant. In VBScript, you can't declare the type since all variables are of the type Variant. An example of a variable declaration is:

```
Dim MyVar
```

Multiple variables can be declared and separated by commas as in:

```
Dim MyVar, YourVar, OurVar
```

To give a variable an initial value, type the following:

```
MyVar = 10
```

The variable has also established the subtype of integer with the value of this assignment. This is because the value assigned is an integer.

Some other things to consider about variables in VBScript:

Variables do not have to be declared in VBScript. You can create a variable by using the variable name. It is a good practice to declare variables as a form of documentation.

What can you name a variable? Almost anything that you wish. Variable naming restrictions are that the name must begin with an alphabetic character, cannot contain an embedded period, must not be longer than 255 characters, and must be unique within the scope of the variable.

OK, what is the scope of a variable? Scope and visibility of a variable describe how long the variable is available and what other parts of the program can see the variable. A variable created in a procedure is limited to the procedure. A variable that is created in the main part of the script can be seen by the entire program or script.

An array is a set of items that are usually the same data type such as a list of names. There can be two-dimensional arrays such as a list of names and phone numbers.

- Arrays may be created as:

  ```
  Dim MyArray (10)
  ```

  This creates an array that can hold ten items.

- Multiple dimension arrays may be created as:

  ```
  Dim MyArray (5, 10)
  ```

  This creates an array with five rows with ten items per row.

Arrays can be resized with the ReDim statement. If the values in the array are to be preserved during the resize, the statement is as follows:

```
ReDim Preserve MyArray (30)
```

VBScript Constants

A constant is a data value that has a name. The value is substituted for the constant in a VBScript statement. The value of the constant never changes. An example of a constant is PI. PI has a value of 3.1416 and on and on. If you are writing a program that uses the value of PI in calculations, after you have declared the constant PI, you can use PI rather than typing out the number. This allows you to determine at the beginning that you will use PI with four decimal places for example and that will be uniform throughout the program.

To declare a constant, the key word CONST is used. An example is:

```
CONST PI
```

After you declare the constant you need to initialize its value. As an example:

```
PI = 3.1416
```

When you are declaring a string constant, the value is enclosed in quotation marks as

```
Const MY_STRING
MY_STRING = "This is my string."
```

For date constants, the value is enclosed in the pound sign (#).

```
Const MY_BIRTHDATE
MY_BIRTHDATE = #5/15/1955#
```

(This isn't actually my date of birth so don't send cards or presents!)

Control of Program Flow in VBScript

There are two basic types of program flow-control statements. These are conditional statements such as the If...Then...Else type of logic and the repetitive execution of code or Loops.

There is one type of program flow-control statement that is present in Visual Basic that does not exist in VBScript. This is the GOTO. The primary issue that the lack of a GOTO presents to the programmer is the requirement to perform all error handling in line.

The conditional statements include the single line If...Then syntax as in:

```
If A = B Then Myfunction
```

and the multiple-line or block If...Then syntax as in:

```
If A = B Then
Myfunction
Else
Anotherfunction
EndIf
```

If statements can be nested just as in Visual Basic and the ElseIF usage is valid in VBScript.

The Select Case statement is also supported as:

```
Select Case MyVariable
     Case Value_1
           Function_1
     Case Value_2
           Function_2
     Case Else
           Function_Else
End Select
```

The select case is very useful when choosing among several alternatives. If the only choices were chocolate and vanilla, you could use logic such as "IF chocolate is available THEN I will eat it ELSE I will eat vanilla." Now when Rocky Road is added to the choices we need the SELECT CASE logic.

For repetitive processing, VBScript supports Do...Loop logic identically to the forms supported in Visual Basic. The Exit Do function is supported and the Do Until and Do While forms work. As an example:

```
Sub MyDoLoopExample()
     Dim mycounter, someNumber
     mycounter = 0
     someNumber = 0
     Do Until someNumber = 100
         someNumber = someNumber + 1
         mycounter = mycounter + 1
         If someNumber > 10 Then Exit Do
     Loop
     MsgBox "The loop made " & mycounter & " repetitions."
 End Sub
```

This loop should make 11 repetitions.

VBScript also supports the While...Wend and the For...Next statements.

VBScript Operators

Operators are used to manipulate variables. For example, the plus (+) operator is used to add two numbers together. The equals (=) operator is used to compare two values.

VBScript has the full range of Visual Basic operators, including logical operators, comparison operators, concatenation operators, and arithmetic operators. When several operators are used in succession, Visual Basic has rules of Operator Precedence that are followed.

- For a list of arithmetic operators, see Table E on the Tear-Out card.
- The comparison operators are used to compare values or objects. The result returned is True or False. The comparison is often used as a test as in "If A = B Then..." (See Table F on the Tear Out card for a list of comparison operators.)
- Logical operators return a value of True or False. As an example, in a test that is selecting rows from a table based on testing the values found in column A and column B, you could use the logical statement "Select the row if Column A = 10 And Column B = 25". Using the And operator, both conditions must be true. A full explanation of the logical operators is beyond the purpose of this chapter. For a complete explanation, the VB 5.0 help files are excellent. (For a list of logical operators, see Table G on the Tear-Out card.)

String Plus String: Concatenation

There is one concatenation operator in VBScript: &. It is used to put two strings together. As an example, if you want to create a new string value that is made up of two other strings, the statement would be:

```
StringVarNew = StringVar1 & StringVar2
```

If StringVar1 contained "ABC" and StringVar2 contained "DEF", then StringVarNew would contain "ABCDEF".

VBScript Procedures

VBScript utilizes two types of procedures, Sub and Function. The basic difference is that a Function can return a value and a Sub cannot.

Both a Sub and a Function can take arguments. The Sub shown in the listing below calculates your weight in kilograms based on an input in pounds.

```
Sub WeightConversion()
     WeightLbs = InputBox("Please enter your weight in pounds.", 1)
     WeightKg = WeightLbs / 2.2
     MsgBox "Your weight in Kilograms is " & WeightKg
End Sub
```

If you wanted to create a Function that would perform the weight conversion, it would be like the next listing shown, which receives the weight in pounds and returns the weight in kilograms.

```
Function ConvertWeight(WeightLbs)
     ConvertedWeight = WeightLbs / 2.2
End Function
```

When this function is used the code is:

```
WeightKg = ConvertWeight(WeightLbs)
```

The Sub or Function can be placed anywhere in the HTML document. A suggested location is in the <HEAD> section of the document. When it is located here it does not affect the appearance of the document and does not have to be placed in comment tags to keep it from printing on a browser that is not VBScript-capable.

VBScript MsgBox Warning The MsgBox function cannot be used in Active Server VBScript. This is because the Message Box will not appear on the client systems browser since it cannot be created with HTML.

Coding Conventions

Coding conventions are not mandatory, but if you ever have to maintain code written by someone else, you will wish that they were. Coding conventions are aimed at improving the readability and maintainability of the code (they do not affect the code in any way). Not everyone uses the same conventions. The key is to establish a set of conventions that you use and adhere to; otherwise code that you wrote two weeks ago can look like it was written by a stranger when you come back to it.

The most important coding convention is the naming of objects, variables, constants, and procedures. If you can look at a name in VBScript such as sdblMySomething and tell that it is a variable that is a double precision, floating point number that has script-wide scope and visibility, then the naming convention is working for you. For example, looking at the variable name sdblMySomething, the first s tells that it has script-wide visibility and the dbl tells that it is a double precision floating point number. MySomething is a descriptive name.

Some other issues to keep in mind:

➤ A suggestion for the naming of constants is to make them all uppercase and separate parts with an underscore as in MY_CONSTANT. (See Table H for a list of suggested prefixes of variables based on data subtype.)

- Procedure names are best when they begin with a verb. OpenMyFile conveys action, which is what a procedure does, as would CheckValues.
- Object names can be challenging since there are so many objects available in VBScript. A few examples will provide some guidance.
    - ckb - CheckBox
    - txt - Text Box
    - cmd - Command Button
- Commenting your code is vital. If you think that you have put in too many comments, you are probably getting close to what you will want six months from now when you look at the code again. Some ideas for things that should be commented:
    - Comment variables when they are declared.
    - Comment procedures. What is its purpose, what are the arguments, what is the return value, where is it used?
    - The beginning of Script comments. They should lay out the purpose of the script and any assumptions that were made.

  If you answer the questions of who, what, where, when, why, and how in your comments, you will have good comments.
- Finally, use indenting and white space to set off sections of code that go together. It will improve readability greatly.

Because most VBScripts are not long, involved programs, I have noticed a tendency to ignore comments and coding conventions. It will come back to haunt you; at least, that is always the case with me.

Using Forms with VBScript

When working with Active Server, you want to maintain independence of the features supported by the Web browser. In order to create interactive Web pages, you must have the user input information from the client system. HTML forms are the method available for collecting information from the user at the client system that does not require any features beyond HTML.

In order to understand the process that is at work here, you will need to create two HTML pages. These pages should be saved in a Virtual Directory on your Web Server. (If you need help refer to Chapter 24, "Creating Your First Sample Project: Where Are the Blueprints," in the section on the functioning of a server.)

The first HTML page will be named CIG_DemoForm.HTM. This will be the form that is used to gather the information from the user on the client system. The code for this file is shown in the listing below. This HTML page will be used to gather the information needed by the Active Server page.

```
<HTML>
<HEAD>
<TITLE>CIG Active Server Demonstration Request Form</TITLE>
</HEAD>
<BODY BGCOLOR = "#FFFFFF">
<H1>Demonstration Input Form</H1>
<H2>This Demonstration Will Calculate Your Approximate Age In Months</H2>
<PRE>
<FORM METHOD=POST ACTION="CIG_DemoForm.ASP">
<P><B> Your First Name:      <INPUT TYPE="text" NAME="fNameVar" SIZE=40>
<P><B>  Your Last Name:      <INPUT TYPE="text" NAME="lNameVar" SIZE=40>
<P><B>        Your Age:      <INPUT TYPE="text" NAME="AgeVar"  SIZE=10>
</PRE>
<INPUT TYPE="Submit" VALUE="Do It"><INPUT
TYPE="Reset" VALUE="Clear">
</BODY>
</HTML>
```

Check This Out...

Use the % Sign All VBScript is enclosed in percentage signs: %VBScript Code%, for example. This lets Active Server or the Browser that is reading and interpreting the HTML page know that it is VBScript, not text.

The second HTML page is the Active Server page that will perform the processing based on the information provided by the input form. The code for this page is shown in the listing below. This form contains VBScript that will process the data returned by the HTML form. Save this form in the same virtual directory with the name CIG_DemoForm.ASP.

```
<%@ LANGUAGE="VBSCRIPT" %>
<HTML>
<HEAD>
<TITLE>CIG Acitve Server Response Form Demonstration</TITLE>
</HEAD>
<BODY BGCOLOR = "#FFFFFF">
<H1>CIG Demonstration Response Form</H1>

<%firstname = Request.Form("fNameVar")
  lastname  = Request.Form("lNameVar")
```

```
  ageyear =   Request.Form("AgeVar")
  agemo = ageyear * 12%>
<%Response.Write(firstname & " " & lastname)%

Your Age In Months Is
<% Response.Write(CStr(agemo))%>

</BODY>
</HTML>
```

Before you examine the code to see what is happening, you should look at the pages with a browser. Open your Web browser and set the URL to **http://systemname/CIG_Demo/CIG_DemoForm.HTM**. You will see the Web page shown in the figure below.

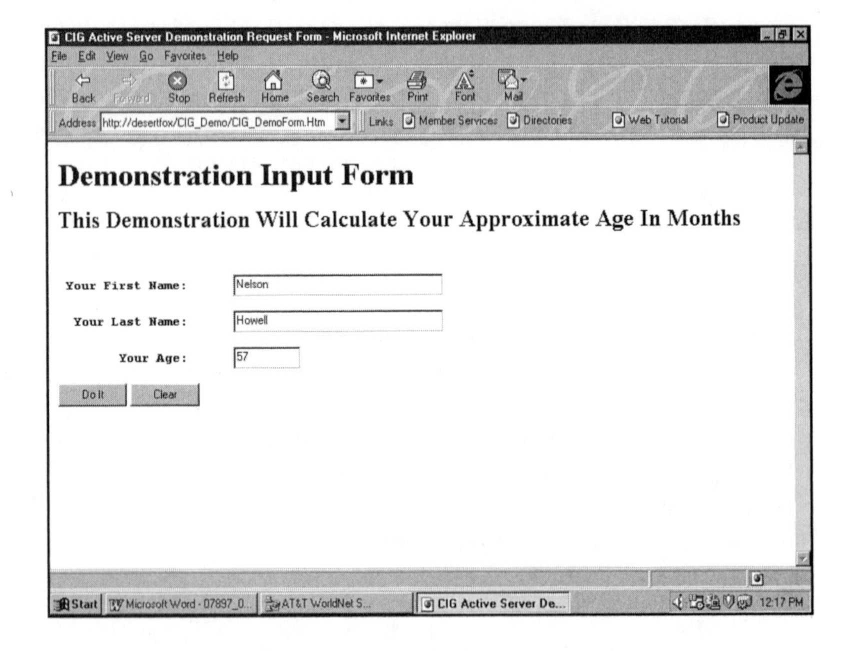

The data collection is performed by this HTML form shown here with sample data entered.

After entering your name and age as shown in the figure above, click the command button labeled **Do It**. The data in the text boxes is submitted to the Active Server page and the response as seen in the following figure is returned to the client browser.

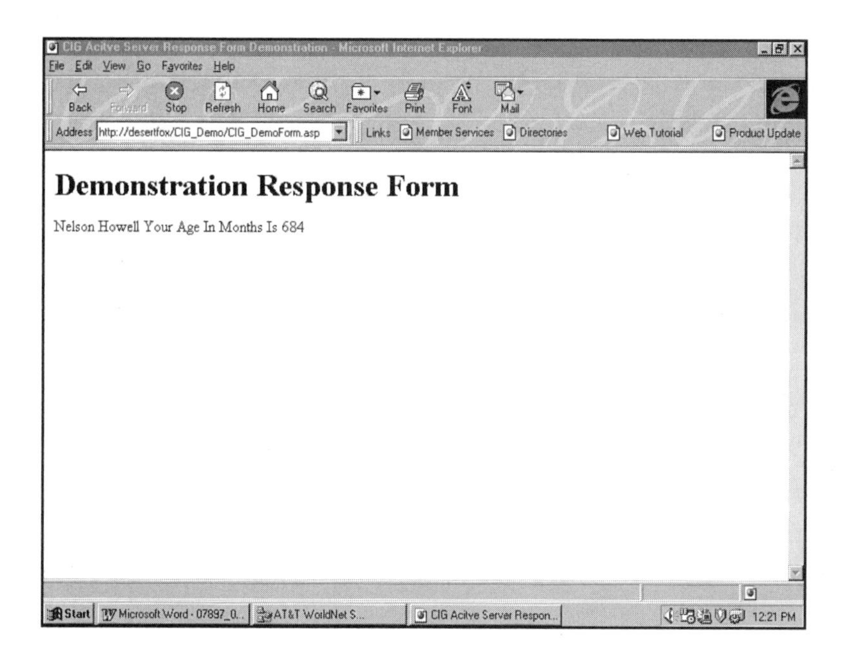

The page returned was created by the Active Server page.

Examining the Code

If the code for the two pages you examine, you will see that it is very simple code to create. There are three key sections in the VB5_DemoForm.HTM listing that deserve further examination. The first is the line of code:

```
<FORM METHOD=POST ACTION="CIG_DemoForm.asp">
```

This code states that the Form Method is to pass the information and the file that it is to be passed to is CIG_DemoForm.ASP.

The next lines of code that are key are as follows:

```
<P><B> Your First Name:      <INPUT TYPE="text" NAME="fNameVar" SIZE=40>
<P><B>  Your Last Name:      <INPUT TYPE="text" NAME="lNameVar" SIZE=40>
<P><B>        Your Age:      <INPUT TYPE="text" NAME="AgeVar"  SIZE=10>
```

These lines create the input test boxes that are used for the data entry.

Next is the command button that submits the data:

```
<INPUT TYPE="Submit" VALUE="Do It"><INPUT TYPE="Reset" VALUE="Clear">
```

The command button created is a “SUBMIT” button that has only one purpose, it initiates the process of sending the request to the .ASP document.

When you examine the code in the Active Server page, you again find three critical sections. The first moves the variables from the input form variables to the Active Server VBScript variables as shown below:

```
<%firstname = Request.Form("fNameVar")
lastname  = Request.Form("lNameVar")
ageyear =   Request.Form("AgeVar")
```

Active Server Pages versus CGI If you have had any exposure to CGI, you will see that the process is somewhat similar. There is one striking difference and that is the ease of creation and maintenance of the VBScript compared to CGI in C++ . There is no compilation required for VBScript.

The next line performs the calculation:

```
agemo = ageyear * 12%>
```

The next section of code writes the data to the response HTML form that is passed to the client browser:

```
<%Response.Write(firstname & " " & lastname)%>
Your Age In Months Is
<% Response.Write(CStr(agemo))%>
```

The Least You Need to Know

The only way to learn a programming language is to write programs. VBScript is very easy to learn and use. Don't be discouraged if at first you stumble over program logic. Programmers are made, not born. Don't think that you will have made the first mistake. (In programming, mistakes are called bugs. Have you ever noticed that bugs outnumber people? There are plenty around.) Dig through samples of programs that others have written—it is a great way to learn. Personally, I wouldn't trade a truck load of documentation for a good working sample program.

Don't assume that because VBScript is easy to learn that it is not a powerful language. The limitations are few and far between. You can do almost anything that you need to in VBScript. Just remember:

- When you are writing VBScript for Active Server, declare the script language with <%@ LANGUAGE="VBSCRIPT" %>. This is because VBScript isn't the only script language that you can use.
- Enclose your VBScript in the <%   %> tags so that the Active Server will know that it is script and not just text.
- HTML forms are the method that is available to collect information from the client user.
- When you are using VBScript in .ASP HTML files, you are not concerned with the features supported by the client browser.
- Choose and use coding conventions. The boogeyman will get you if you don't.

Chapter 9

Java, JavaScript, Java Applets: More Than Just a Cup of Coffee

In This Chapter

- Examine the features of the Java language
- Explore the relationship of JScript to Java
- Create a Web page with JScript
- Learn program flow control with JScript

Java, in times past, denoted either a country in Southeast Asia or slang for coffee. Now it also represents a whole array of technology that was originated at Sun Microsystems. Java the language (Visual J++ is the Microsoft implementation), is used to create applets that are widely used on the Internet to provide active and interactive Web content. It is a scripting language that can be embedded in HTML documents and interpreted by Web browsers with a Java Virtual Machine. The scripting language is called JavaScript, with JScript being the Microsoft implementation.

Check This Out...

Check out the CD-ROM! This chapter has corresponding sample files on the included CD-ROM. Simply click on "Examples," then the corresponding chapter number for which you are interested. (Be sure to read the Readme text file, also located in the "Examples" Section for important information on installing the sample files.)

JavaScript vs. JScript JavaScript is licensed by Netscape and JScript is licensed by Microsoft. For all practical purposes, the two are identical.

Java Virtual Machine Java programs are interpreted programs—interpreted by a program called the Java Virtual Machine, written for a specific computer/operating system platform. A Java program will run on any computer with a Java Virtual Machine.

Java the Language (and of Course Visual J++)

Sun Microsystems describes Java as a simple, object-oriented, statically typed, compiled, architecture neutral, multithreaded, garbage collected, robust, secure, extensible, well-understood language. What did you say?—Run that by me one more time! We will attack this mouthful definition in a moment, but first, why do we care?

Let's start by looking at why Java and Visual J++ are so talked about today for use on the Internet. The small Java-program Web page components called applets mean small applications, not small apples. (OK, no more bad jokes.) These Java applets are embedded in a Web page and transferred with the Web page to the client system for execution (much like an ActiveX object).

Java applets cannot do any damage to other components on the client machine. This means that when I allow you to send a Java applet to my computer, I have not invited the equivalent of a virus to visit.

Java Definition in Detail

Now let's look at that Java definition again, part by part:

- **Simple** The creators of Java intentionally omitted many features of other high-level languages that they considered unnecessary.
- **Object-oriented** Java uses classes to organize code into logical modules. At runtime a program creates objects based on the classes.
- **Statically typed** This means that all objects used in the program must be declared before they are used. This allows the Java compiler to detect and report any conflicts.
- **Compiled** Java is both compiled and interpreted. It is compiled into bytecode. Bytecode is much like the machine language that other compiled languages produce. This bytecode is interpreted by the Java Virtual Machine.
- **Architecture neutral** A Java application can run on any computer architecture that has a Java Virtual Machine.

- **Multithreaded** Java programs can have several threads of execution. This allows several tasks to be performed concurrently.
- **Garbage collected** Java does its own garbage collection. The programmer does not have to delete objects when he is finished with them. This process prevents dreaded memory leaks.
- **Robust** The Java Virtual Machine checks all accesses of system resources before they occur. This keeps the system from being crashed by a Java program. When a serious error occurs, the Java program creates an exception. The exception is dealt with by Java without bringing the system down.
- **Secure** Java checks all memory accesses. This keeps viruses from hitching a ride. Java also doesn't support pointers so there are no accesses to memory outside the area where the Java program is authorized to operate.

> **Memory Leak**
> The much dreaded memory leak is a result of an object being created in memory and not deleted when the program closes. If you open and close a program with a memory leak enough times, the computer runs out of available memory.

- **Extensible** Java supports functions written in another language. These are called Native Methods. This can actually defeat some of the Java features since the function, if written in C++ for example, will be platform-specific. It is expected that as Java matures, Native Methods will not be needed.
- **Well-understood** Java syntax looks much like C++ which is broadly used.

Java Applets

Java applets are small programs that can easily be included in an HTML page. These programs are small enough that transferring them from the host to the client system where they run is quick. The HTML tag that is required to include an applet is very simple:

```
<APPLET
ALIGN=LEFT|CENTER|RIGHT
CODE=appletFile
CODEBASE=codebaseURL
</APPLET>
```

This is an example of the minimum that is required. Java applets are readily available, can be used to provide objects for your Web page, and can be used with JavaScript.

JScript: Microsoft's Implementation of JavaScript

JScript Advantage
The advantage that JScript has at present is the support of JScript by the Netscape Navigator Web browser.

JScript is not Java the language. The two are only distantly related. The two have only some syntax features in common.

The reasons for the use of JScript are the same as were discussed in Chapter 8, "VBScript: Lights, Camera, Action." The use of JScript is an alternative to the use of VBScript, and your choice will depend more on your preferences than capability— both script languages are very capable. (JScript can be used successfully with Active Server pages and any other place VBScript is used.)

Writing JScript Code

JScript code statements consist of one or more items and symbols on a line. A new line begins a new statement, but it is a good idea to terminate your statements explicitly. The semicolon (;) is the termination character:

```
MyVar = "Hello World";
```

When you surround JScript statements with braces, you create a block. Blocks of statements are used in functions and conditional statements. As an example:

Learning JScript
The amount of information that we are able to present on the subject of JScript can only whet your appetite. The documentation that comes with Visual InterDev is very good in the area of JScript, but if you are not familiar with programming, I would recommend that you learn with a good book on JScript or JavaScript, like the *Complete Idiot's Guide to JavaScript, Second Edition,* which is included on the CD-ROM at the back of this book!

```
function convertdistance(Yards)
{
    Miles = Yards / 1760;
    Feet = Yards * 3;
}
```

These two statements are in a block.

Some other points to consider:

- To add comments to JScript, a single line begins with //.
- Multiline comments begin with a forward slash and asterisk in combination (/*), and end with the reverse (*/).
- Assignments in JScript are made with the equal sign (=).

Variables

Variables store values in your scripts. Variable names must adhere to certain format rules:

- The first character must be a letter (either upper- or lowercase) or an underscore ("_").
- Subsequent characters can be letters, numbers, or underscores.
- Reserved words are not allowed.
- JScript is case-sensitive. The variable myVariable is not the same as MYVariable.

Declaring variables is not required, but it is considered good practice to declare variables before using them. You do this with the var statement. (You must use the var statement when declaring variables that are local to a function.) At other times, using the var statement to declare variables before their use is optional.

Examples of variable declaration:

- var MyString = "Hello World!!"; This is a string.
- var MyNumber = 3; This is a numeric type.
- var MyBool = true; This is a Boolean type.

Under some circumstances, you can force the automatic conversion (or coercion) of a variable or a piece of data into a different type. Numbers can easily be coerced into strings, but strings cannot be included directly in numbers. Because of this, explicit conversion functions, parseInt() and parseFloat(), are provided. An example of Coercion is:

```
var MyStart = 1;
var MyEnd = 10;
var MyString = "Count from ";
MyString += MyStart + " to " + MyEnd + ".";
```

When this code executes, the MyString variable contains "Count from 1 to 10." The number data have been coerced into string form.

Data Types

JScript uses five data types. The types covered here are numbers, strings, and Booleans. The others types, functions and methods, as well as arrays and objects, are covered in other sections of this chapter.

- **Strings** Are also called string literals. These are a string of characters enclosed in single or double quotes. Example, "Hi there".

- **Numbers** All numbers are one data type in JScript. There are several types of numbers such as integers, floating point numbers, hexadecimal numbers, and octal numbers. Refer to the Visual InterDev documentation on JScript for a detailed discussion.
- **Boolean** The only Boolean values are true and false. These are not equal to 1 and 0. They are special values.

Operators

Operators are used to manipulate variables. For example the + operator is used to add two numbers together. The == operator is used to compare two values.

JScript has a wide range of operators, including computational operators, logical operators, bitwise operators, and assignment operators. When several operators are used in succession, JScript has rules of Operator Precedence that are followed.

- **The logical operators** return a value of True or False. The comparison is often used as a test as in "If A == B Then..."
- **Assignment operators** are used to assign a value to a variable. For example Myvar = 4 assigns the value of 4 to the variable Myvar. There are also compound assignment operators. An example is Yourvar += Avar would add the value of Yourvar to Avar and assign the result to Yourvar. The compound assignment operators are also used to concatenate two strings. Consult the JScript documentation for a detailed description of all the possibilities of the compound assignment operators.
- **The bitwise operators** are somewhat complex and would require more space than we have to provide an adequate explanation. Unless you are an experienced C++ or J++ programmer, you probably won't use the bitwise operators. If you are interested, the Visual InterDev documentation on JScript is a good place to start.

Program Flow

The simplest program flow statements are the if statement and the if...else. These are called conditional statements. In the if statement a condition is tested and if it is true, the next statement is executed. If the condition is false the statement is skipped. In the if...else, if test 1 is true, statement 1 is executed, if false, statement 2 is executed.

```
If time == morning greeting = "Good Morning";
```

JScript supports several types of repetitive execution or looping.

- **for Loops** This loop has a counter that counts the repetitions and when the target count is reached, the loop is exited.
- **for...in Loops** This loop steps through all of the properties in an object of array. The counter is the number of properties in the array.
- **while Loops** The while loop does not have a counter, it has a condition. When the condition is met the loop is exited.

JScript has the break and continue statements that are used to stop execution of a loop upon a condition and to skip part of the execution of the loop.

Functions

Multiple operations are combined under one name in a function. You can write out a set of statements once, name it, and execute it any time you want to, calling it and passing to it any required arguments. Arguments are passed to a function by enclosing them in parentheses after the name of the function. Functions may take many, one, or no arguments.

JScript has two types of functions. Some are built into the language. Others you create yourself.

An example of a special built-in functions is

```
eval(aString)
```

The eval( ) function evaluates any valid mathematical expression that is presented in string form. The eval( ) function takes one argument, the expression to be evaluated.

You can create your own function and use it where you need it. A function definition consists of a function statement and a block of JScript statements.

```
function addnumbers(a, b)
{
     c = a + b;
     return c;
}
```

This function will add the two numbers a and b together and return the sum.

JScript Intrinsic Objects

JScript has several predefined objects. These are the Array, String, Math, and Date objects. These intrinsic objects have associated methods and properties. These are described in detail in the Language Reference in the Visual InterDev documentation. A couple of examples will help you understand these objects.

To create a new array, use the new statement and the Array () constructor, as in the following example.

```
var MyChildren = new Array(4)  {
MyChildren[0] = "Matthew";
MyChildren[1] = "Andrew";
MyChildren[2] = "Christian";
MyChildren[3] = "James";
}
```

If we now have the line of code:

```
MyVar = MyChildren[3]
```

MyVar will contain "James".

Strings are objects in JScript. When you declare a string variable you are creating a new string object. String objects have built-in methods that you can use with your strings. An example of a method is the substring method, which returns part of the string. It takes two numbers as its arguments.

```
MyString = "ABCDEFGHIJ";
var MyPart = MyString.substring(3, 8);
```

MyPart now contains "CDEFGH".

Creating Objects

You can create your own objects in JScript. You must define the object before you can create an instance. You define it by giving it properties and methods. The following example defines a car object. Notice the keyword "this," which you use to refer to the current object.

```
function car( wheels, seats, speed, cost)
{
        this.length = 4;  //  Number of properties in the object, not
including this one.
        this.wheels = wheels;   //  Number of Wheels
        this.seats = seats;   //  Number of seats
        this.speed = speed;     //  How fast will it go(MPH)
        this.cost = cost;    //  What does it cost (dollars)
}
```

Now that we have the object defined, we can create an instance. We use the new statement:

```
var SportsCar = new car(4, 2, 150, 25000);
```

JScript and Active Server

When working with Active Server, one goal is to not transfer anything but HTML to the browser. Creating interactive Web pages will require the user to input information from the client system. HTML forms are the method that you have available to collect information from the user at the client system that does not require any features beyond HTML.

> Check This Out...
>
> **Just Like Chapter 8**
>
> The demonstration in this chapter is as identical to the demonstration project in Chapter 8, "VBScript: Lights, Camera, Action!," as possible so that you will be able to compare the same functions in JScript and VBScript.

To examine the process used we will create a pair of HTML pages. Create a project named jscriptdemo and create a HTML page.

The first HTML page is named CIG_DemoJScript.HTM. This is the form used to collect data from the user. The code for this file is shown below. The information collected will be used by the Active Server page.

```
<HTML>
<HEAD>
<TITLE>CIG Active Server JScript Request Form</TITLE>
</HEAD>
<BODY BGCOLOR = "#FFFFFF">
<H1>Demonstration Input Form</H1>
<H2>This Demonstration Will Calculate Your Approximate Age In Months</H2>
<PRE>
<FORM METHOD=POST ACTION="CIG_DemoJScript.ASP">
<P><B> Your First Name:      <INPUT TYPE="text" NAME="fNameVar" SIZE=40>
<P><B>  Your Last Name:      <INPUT TYPE="text" NAME="lNameVar" SIZE=40>
<P><B>        Your Age:      <INPUT TYPE="text" NAME="AgeVar"  SIZE=10>
</PRE>
<INPUT TYPE="Submit" VALUE="Do It"><INPUT TYPE="Reset" VALUE="Clear">
</BODY>
</HTML>
```

This HTML form is virtually identical to one that would be used with VBScript.

The second HTML page is the Active Server page that will perform the processing based on the information provided by the input form. The code for this page is shown in the listing below. This form contains JScript that will process the data returned by the HTML form. This form is named CIG_DemoJScript.ASP. Create this page in the same project.

```
<%@ LANGUAGE="JSCRIPT" %>

<HTML>
<HEAD>
<TITLE>CIG Active Server JScript Response Form</TITLE>
</HEAD>
<BODY BGCOLOR = "#FFFFFF">
<H1>CIG JScript Response Form</H1>
<%
{
firstname = Request.Form("fNameVar");
  lastname  = Request.Form("lNameVar");
  ageyear =   Request.Form("AgeVar");
  agemo = ageyear * 12;
  }
  %>
<%
{
Response.Write(firstname + " " + lastname);
}
%>
Your Age In Months Is
<% Response.Write(agemo);%>
</BODY>
</HTML>
```

Before you examine the code to see what is happening, you should look at the pages with the browser. Right-click the file name in the Workspace window and choose **Preview in Browser**. You will see the Web page shown in the following figure.

After entering your name and age as shown in the following figure, click the command button labeled **Do It**. The data in the text boxes are posted to the Active Server page and the response in the following figure is returned to the browser.

When you examine this code, refer back to the code in Chapter 8. There are the same three key sections in the CIG_DemoJScript.HTM listing. The first is the line of code:

```
<FORM METHOD=POST ACTION="CIG_DemoJScript.asp">
```

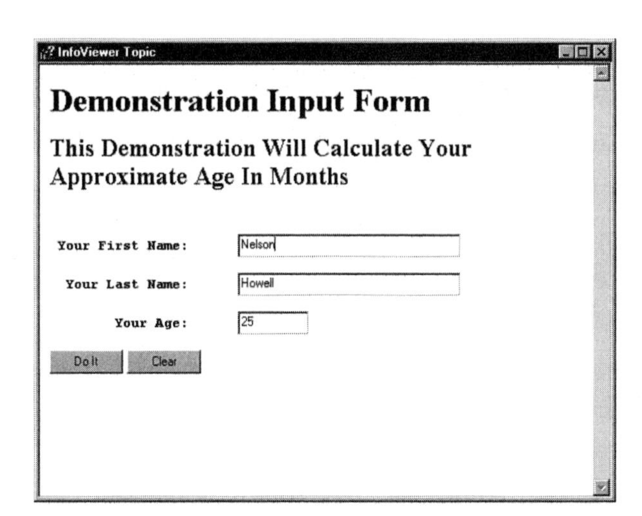

The data collection performed by this HTML form.

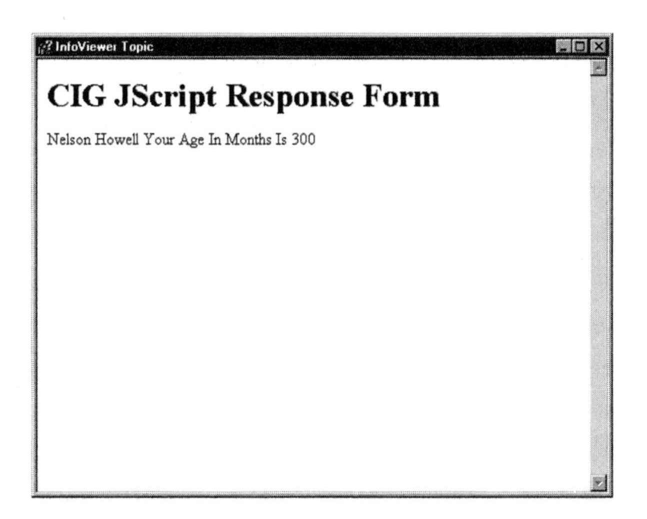

The page returned is created by the Active Server page.

This code states that the Form Method is to pass the information. The file that it is to be passed to is CIG_DemoJScript.ASP.

The next lines of code that are key are:

```
<P><B> Your First Name:      <INPUT TYPE="text" NAME="fNameVar" SIZE=40>
<P><B>  Your Last Name:      <INPUT TYPE="text" NAME="lNameVar" SIZE=40>
<P><B>        Your Age:      <INPUT TYPE="text" NAME="AgeVar"  SIZE=10>
```

These lines create the input text boxes that are used for the data entry.

Next is the command button that submits the data.

```
<INPUT TYPE="Submit" VALUE="Do It"><INPUT TYPE="Reset" VALUE="Clear">
```

The command button created is a "SUBMIT" button that has only one purpose, it initiates the process of sending the request to the .ASP document.

When you examine the code in the Active Server Page, you again find the same three important sections. The first moves the variables from the input form variables to the active server JScript variables as shown below:

```
firstname = Request.Form("fNameVar");
  lastname  = Request.Form("lNameVar");
  ageyear =   Request.Form("AgeVar");
```

The next line:

```
agemo = ageyear * 12;
```

performs the calculation.

The next section of code:

```
{
Response.Write(firstname + "  " + lastname);
}
%>
Your Age In Months Is
<% Response.Write(agemo);%>
```

writes the data to the response HTML form that is passed to the client browser.

When you compare this to the VBScript version, you will see the similarities are greater than the differences.

The Least You Need to Know

JScript is as easy to learn and use as VBScript. Again, the only way to learn a programming language is to write programs. You have seen here that it is very simple, but don't be discouraged if at first you stumble over program logic. Dig through samples of programs that others have written. It is a great way to learn. Personally, I wouldn't trade a truck load of documentation for a good working sample program.

Finally, don't assume that because JScript is simple to learn that it lacks power. The limitations are few, and you can do almost anything that you need to in JScript. Just remember:

➤ When you are writing JScript for Active Server, declare the script language with <%@ LANGUAGE="JScript" %>. JScript isn't the only script language that you can use.

➤ Enclose your JScript in the <% %> tags so that the Active Server will know that it is script and not just text.

➤ HTML forms are the method that is available to collect information from the client user.

➤ When you are using JScript in .ASP HTML files, you need not be concerned with the features supported by the client browser.

Chapter 10

ActiveX Controls, OLE & VB5CCE: Making Your Own

In This Chapter

- Explore ActiveX
- Learn the relationship between ActiveX and OLE
- Insert an ActiveX Control into an HTML page
- Build your own ActiveX Control with Visual Basic 5.0 Control Creation Edition

ActiveX is one of the hottest new technologies for the Internet, and is leading the way in creating interactive content and reusable objects for Web content. It is Microsoft's answer to the Sun Microsystems Java family of technologies. (If you aren't excited yet, check your pulse—you may be dead.)

What is ActiveX

ActiveX is derived from Microsoft's OLE (Object Linking and Embedding), which in turn is based on the Microsoft COM (Component Object Model). ActiveX objects are reusable program functions and are based on the same model as OLE. However, one major change was made to OLE objects. They were simplified and made smaller. This is important since ActiveX objects are embedded in Web pages that are transferred from the server to the client. OLE Control objects (usually called OCXs) can be used in the same way as ActiveX objects. However, the problem with this approach is that OCXs are bigger and take too long to transfer.

You May Be Able To Buy What You Need

There are many ActiveX Controls available commercially. Look at a copy of *Visual Basic Programmer Journal* for instance and you will see many advertisements for ActiveX Controls. These controls cover a range of functionality that staggers the imagination. It is much like the VBX and OCX markets.

Ncompass Script Active Plug-in

Ncompass has created a plug-in for Netscape Navigator that will allow Net-Scape Navigator 3.0 and later to support ActiveX objects. You can see for yourself at **http://www.ncompasslabs.com**.

How an ActiveX Control Moves to the Browser

When the Web page containing an ActiveX Control is transferred to the client browser, the browser looks for the ActiveX Control on the Client system. If ActiveX is present, it is used. If it is not present, the browser sends a message back to the server requesting the ActiveX Control. It is then transferred to the Client system.

Like OLE, the receiving system must be able to host the object for the ActiveX object to perform. Translation: the browser must be ActiveX-enabled. At present, Microsoft Internet Explorer is the only browser able to work with ActiveX, but the soon-to-be-released version of Netscape Communicator will also offer ActiveX support (it may have already been released by the time you read this).

Inserting an ActiveX Control into a Web Page

Using Visual InterDev, the process of inserting an ActiveX Control into a Web page is quite simple. The first task is to create a new project, so open Visual InterDev and choose **File**, **New** and click the **Projects** tab. Select the Web Project Wizard by highlighting it, enter the Project name **CIG_Chap_11**, select **Create new workspace**, and click the **OK** button.

At the next dialog box click the **Finish** button. You have created a project. The next task is to create an HTML page. Choose **File**, **New** and click the **Files** tab. Select the HTML Page by highlighting it. Enter a file name of **default**, make sure the **Add to project** check box is checked, and click the **OK** button to create the file.

You should be looking at the file default.htm in the text edit window.

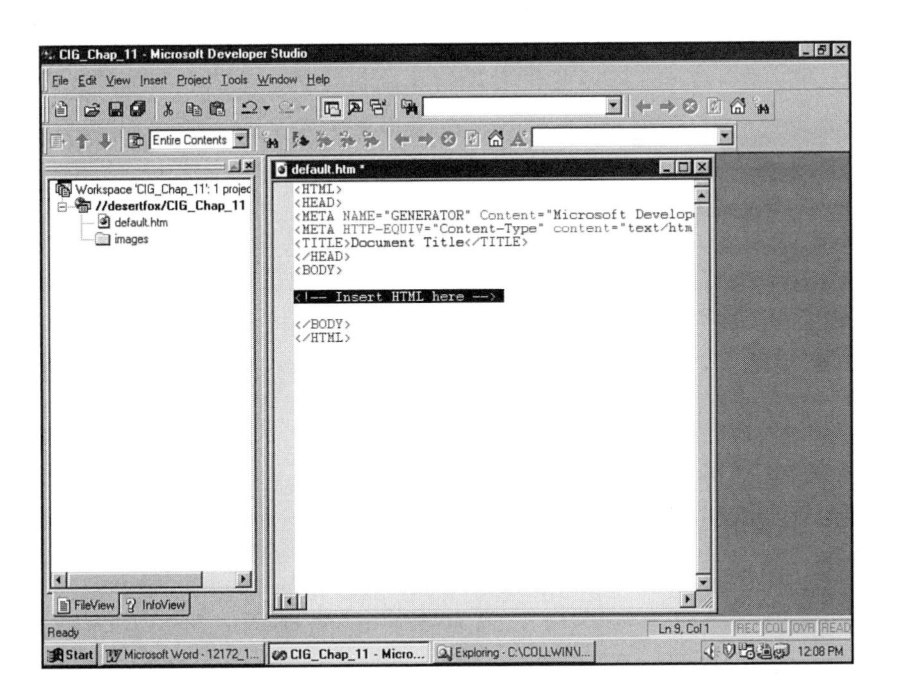

If you were going to use this file in a Web site, you would want to change the "Document Title" to an appropriate name.

You are now ready to begin the process of inserting the ActiveX Control. To begin, right-click the highlighted line of comment code:

```
<!-- Insert HTML here -->
```

Choose **Insert ActiveX Control** from the menu that appears.

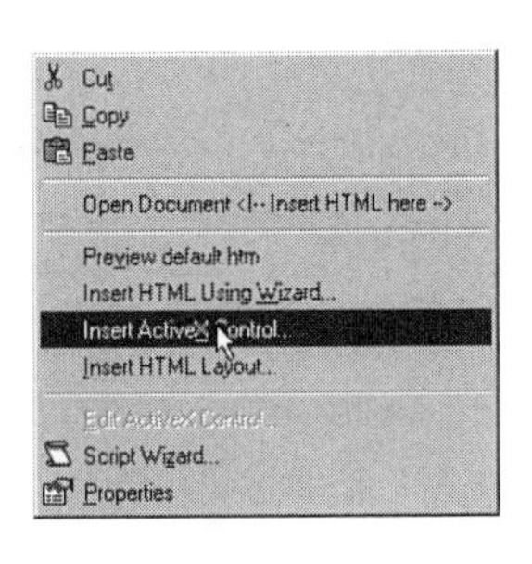

Right-click menus are enabled throughout Visual InterDev.

The Insert ActiveX Control dialog box presents a list of available ActiveX Controls that may be inserted. You select a control by simply highlighting its name.

If you acquire controls from other sources and install them on your system, they will appear in this list.

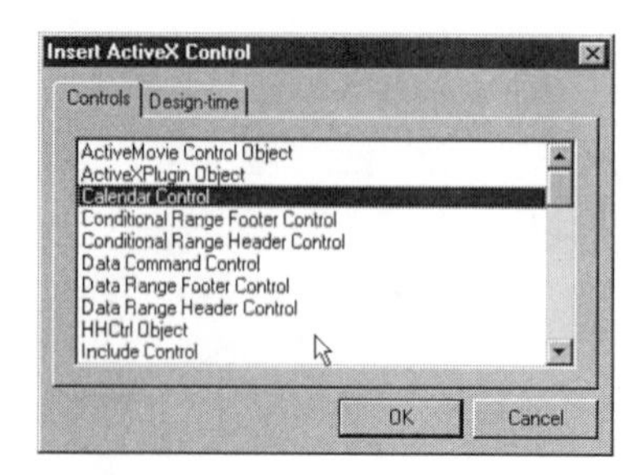

ActiveX Controls Are Everywhere

ActiveX Controls are available from many sources other than Microsoft. If you need a control that performs a particular function, you may find it is for sale. On the Web, you may want to check out **http://www.protoview.com** and **http://www.microhelp.com** as well as Microsoft's site (of course!) at **http://www.microsoft.com/activex/**. Many vendors offer free trial versions for your inspection.

The Calendar Control has been selected for the example. When the **OK** button is clicked, the control appears along with a Properties window. As you can see in the next figure, you can set several options.

The appearance properties of the control can be set here at design time or runtime.

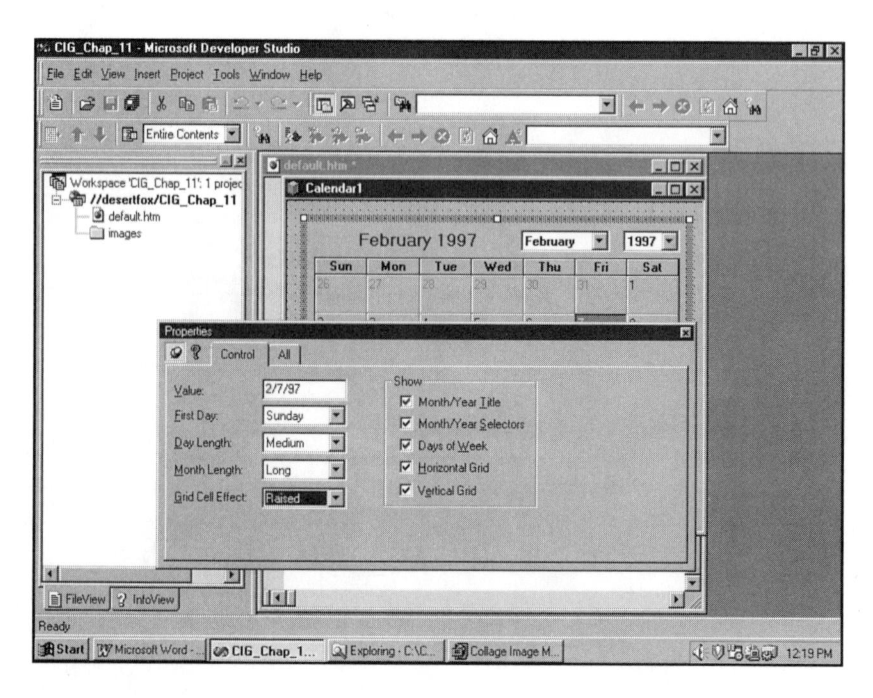

If you click the All tab of the Properties window, you will see all of the properties exposed. A key property you should notice is the ID property. This is the name that you will use to refer to the control and a property in code such as Calendar1.Month. The list of properties as shown in the next figure will vary based on the control.

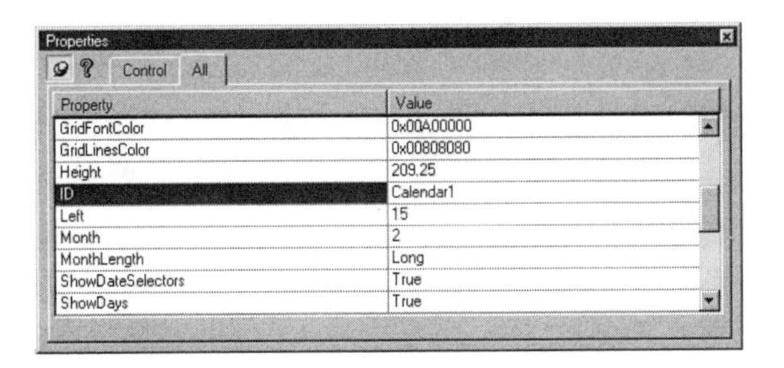

In addition to the ID property, notice the Height property—it determines the height of the control when it is displayed by the browser.

When you close the Properties window and the Control Window, you will see that the <OBJECT> tag has been inserted in the HTML file with the required parameters. In the next figure, the three parameters that determine the date that the calendar opens to are shown as Year = 1997, Month = 2, Day = 7. If these settings remain, the calendar will always open to February 7, 1997. You can alter this to any date desired. By substituting VBScript date functions for the values in the HTML, you can have it open to the current date.

default.htm *

```
<HTML>
<HEAD>
<META NAME="GENERATOR" Content="Microsoft Developer Stud
<META HTTP-EQUIV="Content-Type" content="text/html; char
<TITLE>Document Title</TITLE>
</HEAD>
<BODY>


<OBJECT ID="Calendar1" WIDTH=372 HEIGHT=279
 CLASSID="CLSID:8E27C92B-1264-101C-8A2F-040224009C02">
    <PARAM NAME="_Version" VALUE="458752">
    <PARAM NAME="_ExtentX" VALUE="9843">
    <PARAM NAME="_ExtentY" VALUE="7382">
    <PARAM NAME="_StockProps" VALUE="1">
    <PARAM NAME="BackColor" VALUE="12632256">
    <PARAM NAME="Year" VALUE="1997">
    <PARAM NAME="Month" VALUE="2">
    <PARAM NAME="Day" VALUE="7">
</OBJECT>
</BODY>
</HTML>
```

Experiment with the Month, Day, and Year values to see the effect on the control. Also experiment with the Height and Width properties.

The Big Test

All that remains is for you to test the HTML page to see if your insertion was correct. Right-click the text edit window and choose **Preview default.htm** as shown in the next figure.

You can display a similar menu by right-clicking the file name default.htm in the Workspace window.

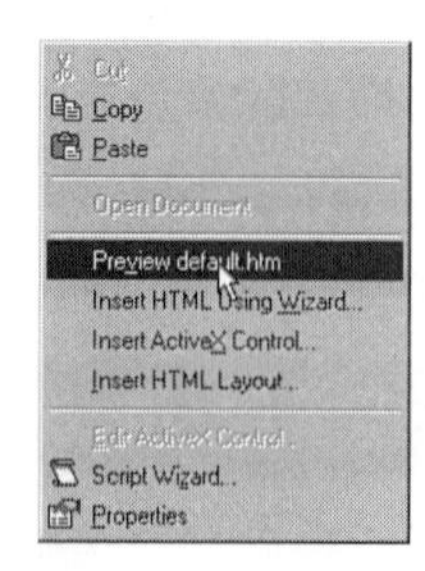

You should be looking at a display in the InfoViewer Topic window such as in the next figure.

The process of inserting an ActiveX Control is very easy and relatively fool-proof. It almost takes concentrated effort to foul it up.

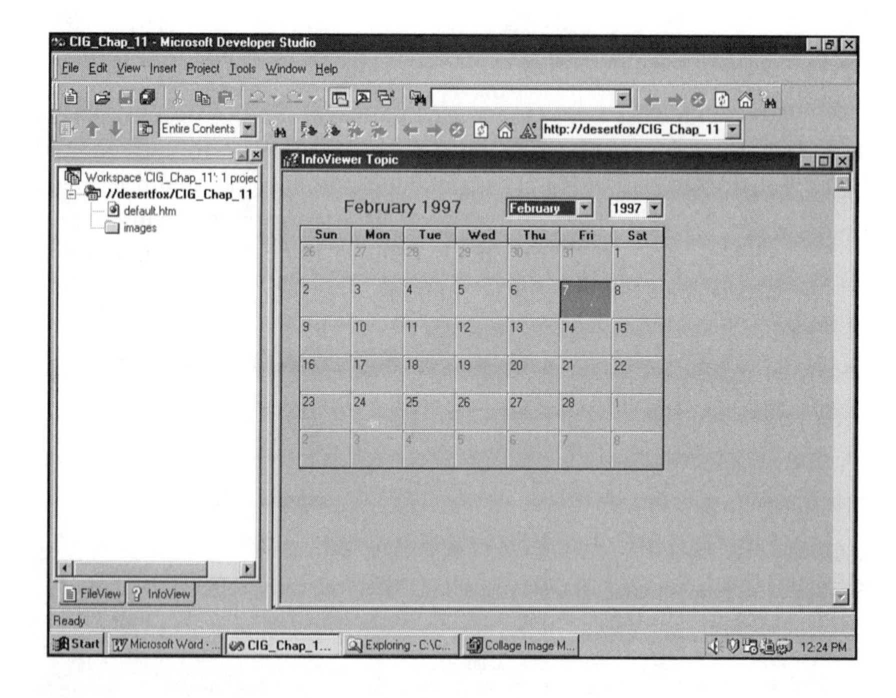

Registering Controls

For reasons of system security, non-Microsoft controls that you use may require Verisign registration and certification. This subject is beyond this book. **http://www.microsoft.com/cio/articles/digital.htm** is a good place to begin researching this technology.

Creating an ActiveX Control With VB 5.0 Control Creation Edition

Even with all of the ActiveX Controls available, there will be times when you can't find exactly what you want to use on your Web page. This is when Visual Basic 5.0 comes to the rescue. The latest version of Visual Basic is capable of creating ActiveX controls as an OCX. In the material that follows, you will see a process that is amazingly simple when you consider the end result.

Visual Basic 5.0

The material in this chapter was created with the Visual Basic 5.0 Control Creation Edition in the Beta release. As this is written, Visual Basic 5.0 Professional Edition has been announced and will be available soon. There will not be any significant changes to the processes that are discussed here. If you don't have VB 5.0, you should read this chapter anyway—it may help you to decide whether or not you want to get it. At this point it appears that VB 5.0 CCE will always be available from the Web at **http://www.microsoft.com**.

Start the task by opening Visual Basic 5.0. You will be presented with a dialog box that offers you the choice of creating an ActiveX control on the New Project dialog. Accept this choice by clicking the **Open** button.

The next step is to name the project "MyProject" and the control "My Control." You are now ready to add a text box and two command buttons to the control MyControl as shown in the figure below. These are the name properties for the project and control respectively.

Next you will double-click one of the command buttons to open a code window as shown in the next figure.

All of the properties of the control and the objects in the control can be displayed in the Properties window on the right.

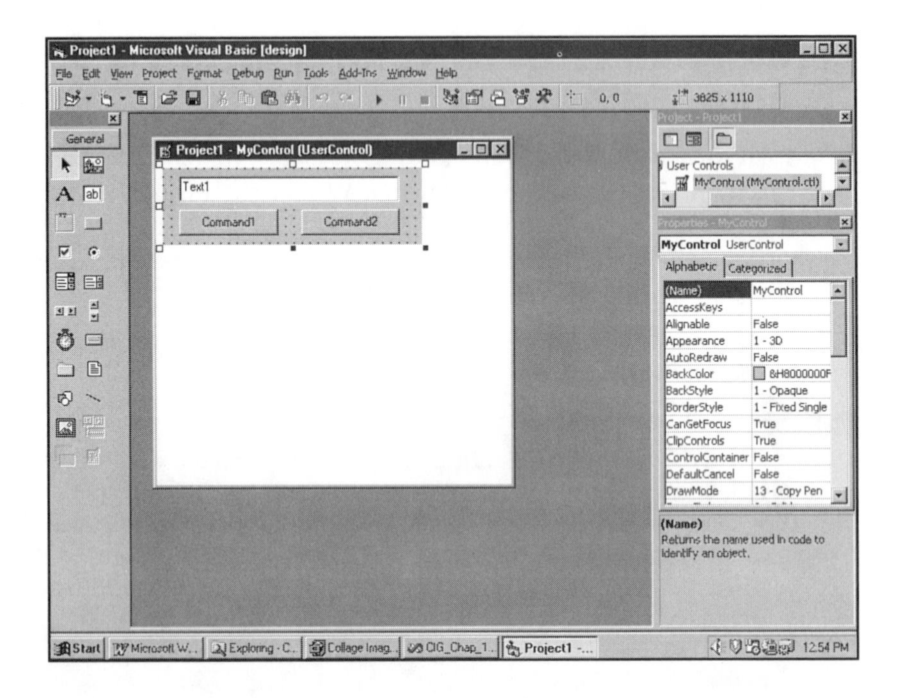

You may add code to any event for any object from this window.

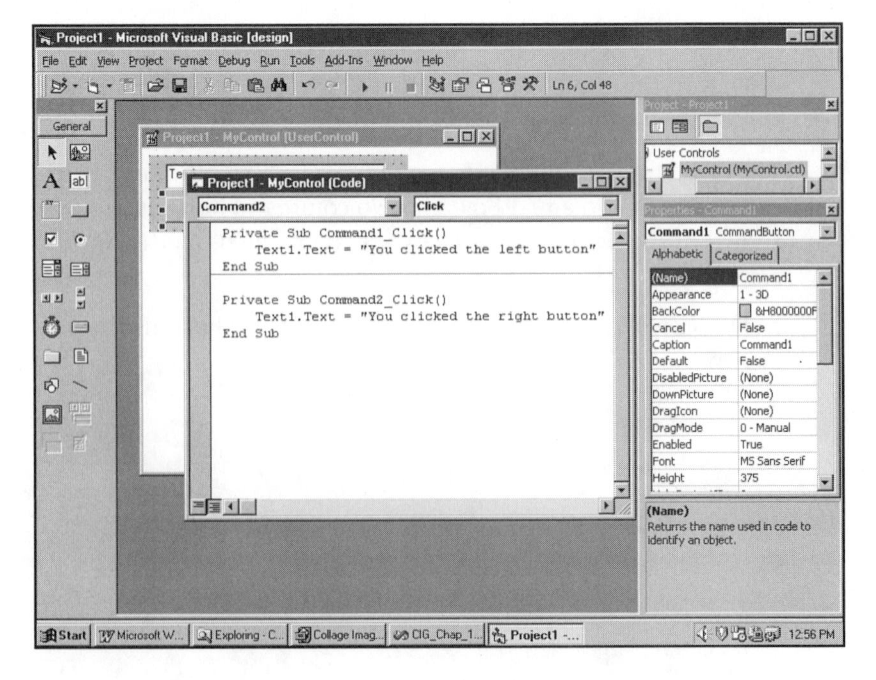

Enter Your Code Please

We are now going to add code to the click event for both of the command buttons. The code simply displays a text string in the text box Text1 indicating that the button was clicked.

```
Text1.Text = "You clicked the left button"
```

This code follows the model of Object.Property = Value that is used in Visual Basic.

> Check This Out...
>
> **More on ActiveX,OCX, and Visual Basic 5.0** For more information on this topic, you may want to check out *Special Edition: Using ActiveX* and *Visual Basic 5 Control Creation Starter Kit*, both published by Que.

Your control is now complete and ready for creation of the OCX. Choose **File**, **Make MyProject.ocx**. A dialog box will open asking where you want the OCX to be saved. Where you save it is not important. However you should choose a location where it will stay because the OCX is registered on the system as part of the creation. You may now close Visual Basic and return to Visual InterDev.

We now have a control that is ready to be inserted into our HTML file. Right-click the default.htm file in the text edit window. This is the same process that was used for inserting a pre-existing ActiveX Control. Find the Control named MyProject.MyControl in the Insert ActiveX Control dialog box as shown in the next figure.

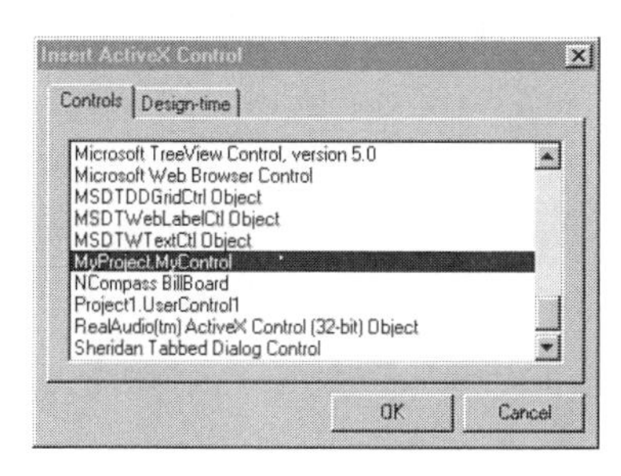

Our systems may become as littered with ActiveX controls as they were with VBXs in Windows 3.1.

It is now time to test your creation. Right-click on the text edit window and choose **Preview default.htm** from the dialog box. When the page is displayed, you will see that both controls are in the page. Now click the command buttons in the newly created AcitveX Control and see the text string displayed in the text box change depending on which button was clicked.

The controls can perform any function that can be programmed in Visual Basic.

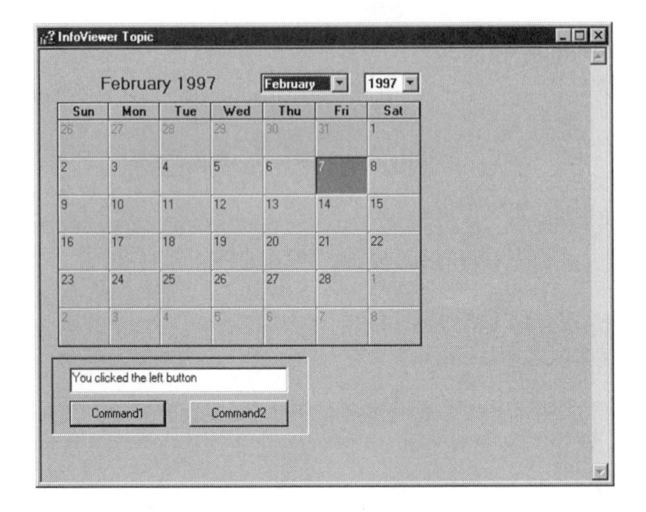

The Least You Need to Know

ActiveX Controls add new dynamic dimensions to Web content. Visual Basic 5.0 also enables you to create your own ActiveX Controls. The sky is the limit.

- Review the ActiveX Control property settings in the Properties window. You can adjust them to your preferences to blend with the remainder of your Web page.
- You can set properties at run-time using VBScript variables to set values for the properties.
- You can create your own controls using Visual Basic 5.0. Your imagination is the only limit to the capability that you can build into a control.

Chapter 11

Multi-Tier Applications

In This Chapter

- Explore the flexibility in multi-tier applications
- Examine how DCOM and Active Server fit into the picture

The present trend in application structure is for different components to operate on different, specialized-use systems. Examples of this are the database servers and the client/server applications. By their very nature, these applications are distributed with different processes taking place on different systems. This logic has been extended in the Object Linking and Embedding (OLE) components development, so we now have a client taking advantage of the methods and properties of a component that are exposed through OLE.

Multi-Tier Applications

A multi-tier application is composed of layers or tiers. An example might be an application with a user interface layer running on a group of systems. The UI could be Web browsers. The next layer is one or more systems in a Web server layer that are processing the interface to the UI layer and communication with a third layer that is composed of database servers.

COM, DCOM, LANS, WANS, and Other Confusing Acronyms

This logic is being extended further with the ActiveX objects that are the latest incarnation of OLE (which is derived from the standard Componet Object Model, or COM). In the figure below, we see a client communicating directly with a component.

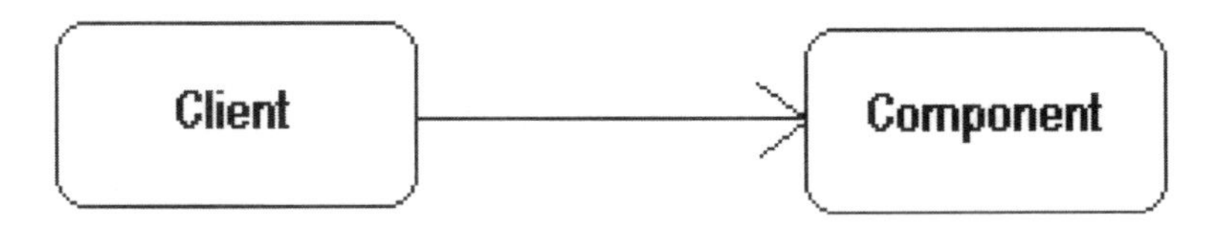

The client communicates with the component's exposed properties and methods.

The modern operating system precludes this type of direct communication between processes. Now the client communicates with the process, which uses interprocess communication that communicates with another process, which in turn communicates with the component.

Did you get all that? The figure below provides a graphic view that should clarify matters.

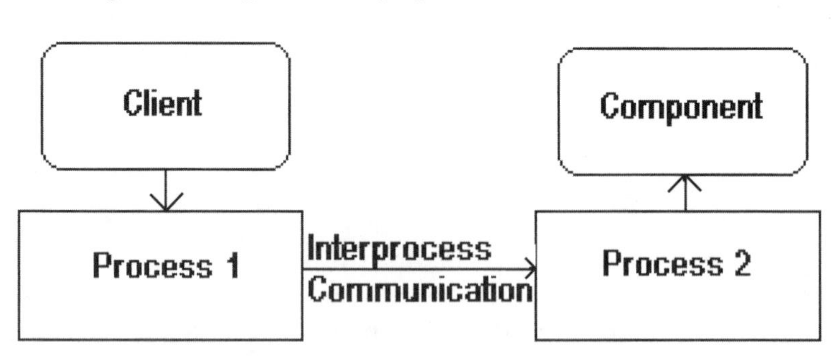

The communication is transparent to the client and the component.

The distribution now can take place on different systems that communicate over a Local Area Netowrk (LAN) or Wide Area Network (WAN). COM has now evolved to the Distributed Component Object Model (DCOM). This means that the processes can be running on different machines. (The client and the component communication is transparent to both.) All that has been added is that the communication is now through a DCOM network protocol and Protocol Stack (the set of programs that handle the interface between the network medium such as an Ethernet cable and the application) as shown in the next figure.

How does this make life easier for developers? It eliminates the management of communication when using a DCOM connection. All of the communication is handled by the operating system transparent to the programmer.

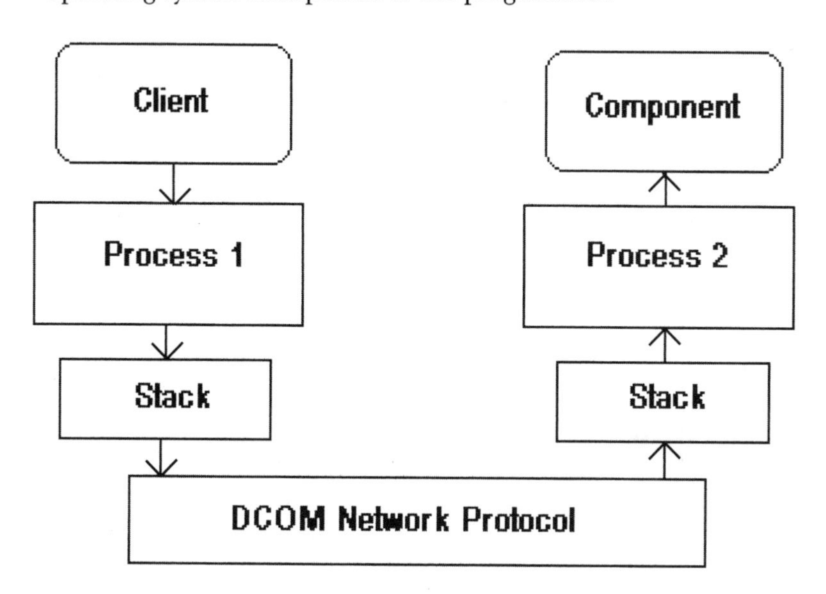

The addition of the DCOM protocol to the communication is transparent to the client and the component.

Platform Independence

DCOM is an open standard that is expected to be available in early 1997 for UNIX and Macintosh, in addition to being part of the Windows NT 4.0 operating system and a component to be added to Windows 95. There are indications that the UNIX and Macintosh availability might slip to a later date.

Distribution for Load Balancing and Application Scaling

If a particular process in an application is providing a bottleneck to an application, the process that is the throat can usually be subdivided into multiple smaller components. With DCOM, these components can then be distributed over multiple systems for load balancing. Also as the use of an application grows, additional systems can be added that perform the "bottleneck" function. These functions can be designed and constructed in tiers that then can be scaled. The following figure shows an illustration of this approach.

DCOM renders the components more independent of location.

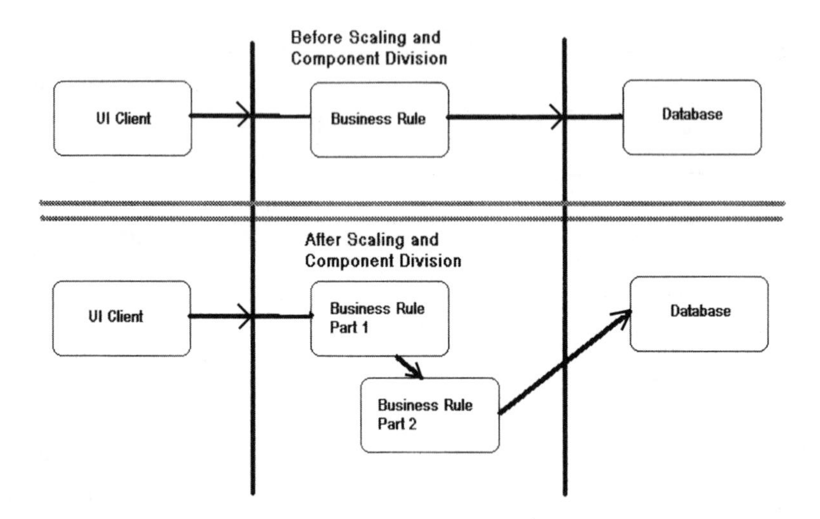

Language Independence

Because DCOM is an extension of COM, it is language-independent. Virtually any language can be used to create COM Components and those components (once created) can be used from many different languages. Some of the languages that interact well with COM and DCOM are MS Visual Basic, MS Visual C++, Java, J++, Delphi, PowerBuilder, and Micro Focus COBOL. This provides the developer with more available choices.

Visual InterDev and Active Server

Active Server is an important feature of Visual InterDev, since Active Server is a COM component and can be driven by VBScript and JScript. The great advantage to Active Server is the fact that it executes in a controlled, known environment on the server. Since it does not execute on the client desktop system, the code is secure and not revealed to the client.

Active Server can easily be used to develop multi-tier applications. It also has the advantage of being able to use most Web browsers as the User Interface (UI) component of the application. Using the ActiveX Data Object (ADO) database components to communicate through Open Database Connectivity (ODBC) to back end databases, Active Server will be able to incorporate many legacy database systems into new applications.

Using the ADO components, you can mine the large corporate databases for data that can then be warehoused on middle tier systems in support of various OLAP (Online Analysis Programs) such as financial modeling. Another example is extracting performance data from sales data and analyzing inventory performance.

The Least You Need to Know

DCOM and multi-tier applications are a rapidly developing area of technology. It is impossible to predict the future in these areas, but some broad outlines are becoming clear. (The trend toward distributing applications will most certainly continue.) Microsoft is developing additional technology for the clustering of WinTel architecture machines, bringing even more processing power to bear on application solutions.

- As the old advertisements said, "Watch this space." The developments in the area of Active Server and DCOM will be coming rapidly and will have significant impact on the way that we do computing.
- The primary limiting factor in the development of the multi-tier applications has been removed with the advent of DCOM and Active Server. No longer will the difficulty of creating custom communication elements be a requirement for multi-tier applications.

Part 4
Databases

Adding database support to Web pages is one of the most significant extensions that has occurred. Relational databases are the most popular storage format for data at the present time and will likely be so far into the future.

In this section, you will explore relational databases, SQL, and the database tools provided by Microsoft Visual InterDev. These tools take the mystery out of database usage.

BARRY WEIN MICROWAVES A BURRITO AND KNOCKS OUT THE HOME OFFICE.

Chapter 12

Relational Databases and SQL: Information, Please!

In This Chapter

- Create a relational database table
- Understand the function of a primary key in a table
- Describe the purpose of a foreign key in a table
- Explain what is meant by normalization
- Understand the clauses in an SQL statement
- Explain the function of the primary key and foreign key in a join
- Write an Add, Update, Delete, and Select SQL Statement

Many view relational databases as an arcane art whose practice is a form of sorcery. Nothing could be further from the truth (he said as he added eye of newt to the cauldron). Relational databases are marvelously logical. Understanding relational databases and SQL is like riding a bicycle. Once you learn, you never forget. (Or is it possible that I am possessed and don't know it? Nah, not me. Pass the tongue of frog, please.)

Warning!

The following material may be skipped if you are already familiar with relational databases and SQL. For those of you who choose to plunge on, the going gets somewhat rough. I have endeavored to keep it brief and not attempt to explain the theory. However, I do consider it essential that you have at least a speaking acquaintance with Relational Database Management Systems (RDBMS) and SQL to be able to work with the database tools that are part of Visual InterDev. With that said, off we go into the wondrous land of Relational Database Management Systems.

Come Join Me at My Table

Tables are the foundation of relational databases. All data is arranged in rows and columns much like a train schedule or a spread sheet. Each row is a record and each column represents a particular item of data. An example is a listing of names and phone numbers of employees. Below is a table showing three rows and four columns. Each row represents one individual and each column represents an item of information about the individual. The top row contains the column names.

First Name	Last Name	Area Code	Phone Number
Sam	Jones	202	555-1212
Tom	Smith	212	555-1345
Dorothy	Rubyslippers	316	444-4321

We may need more than one table in our database. For example, assume that we also need the birth dates of these individuals. We can create another table like the one below.

First Name	Last Name	Birth Date
Sam	Jones	05-25-1955
Tom	Smith	07-14-1964
Dorothy	Rubyslippers	08-29-1981

A reasonable question for you to ask at this point is why we didn't just put the date of birth in the first table. When you consider the issue, you will realize that it is not unusual for a person to have more than one phone number. But each individual should have only one date of birth. That would mean that there could be multiple rows for Sam Jones in the first table but not in the second table. To illustrate the point, the table below has more than one phone number for one of the individuals.

First Name	Last Name	Area Code	Phone Number
Sam	Jones	202	555-1212
Tom	Smith	212	555-1345
Sam	Jones	202	444-9876
Dorothy	Rubyslippers	316	444-4321

It is also possible that we are making an invalid assumption that both Sam Joneses are the same person. If we had the Social Security Number we would be sure.

So, other information that we will need for our employees include: Social Security Number, job title, and rate of pay. In this company, the rate of pay for a given job title will always be the same. All programmers are paid the same hourly rate as are all secretaries, computer operators, and so on.

We can now begin to create meaningful tables that meet the rules of normalization. What's normalization? Please read on.

Normalization

Properly designed databases will follow the rules of normalization. There are three of these rules that are applied to our database design. When a database complies with the rules, it is said to be in the *normal form*. The following sections discuss each of the three rules, known as (surprise) the First Normal Form, the Second Normal Form, and the Third Formal Form.

> Check This Out...
>
> **Primary Key**
> Each table needs to have a primary key. This is a column or group of columns that contains unique data. Two table entries will not share the same primary key. (A phone number is not a good primary key. A Social Security numer is.) The primary key can also be a number that is assigned only for use in the database.

First Normal Form

A table is in the first normal form if all columns in the table contain atomic values. A quick translation of this is that a given column will contain only one phone number for instance, never two phone numbers.

Second Normal Form

A table is in the second normal form if it complies with the first normal form, every column is dependent upon the primary key value, and each row is unique. For example, earlier we had a table where Sam Jones' name appeared twice. If we were attempting to use Sam Jones' name as the primary key, the table doesn't fit the second normal form. The phone number is the only unique data in the table. If we use the phone number as the primary key, the table might be in the third normal form. But you say, people aren't usually identified by their phone number. And can't two people have the same phone number? Now you begin to see that the table design for that table is flawed.

Third Normal Form

A table is in the third normal form if it complies with the second normal form and also if every value in the table depends on only the primary key. Here we can look at adding the pay rate to the table with the names. The pay rate depends on the job title, not the person in the job. Therefore, the pay rate should be in a table with the job title.

Constructing the Normalized Database

Now we will construct our normalized database. You will see the effects of normalization.

The data that we have to work with are:

- Area Code
- Date of Birth
- First Name
- Job Title
- Last Name
- Pay Rate
- Phone Number
- Social Security Number

When you start to design a database, it is good to identify the central item with which the database is concerned. In this case, you will agree that employees are the central item or thing. We will call the first table the People Table.

The People Table

Now we will identify all of the data items about a person in our list of data elements. Each individual has only one. These items are:

- Date of Birth
- First Name
- Job Title
- Last Name
- Social Security Number

Each person has only one pay rate, but remember we said that pay rate is tied to job title (not the individual), so it doesn't belong in this table. Our table is in the first normal form since the values are atomic. Now, we need to select a data element that is unique to serve as the primary key. There is more than one Sam. There is more than one person born on any given day. The one item in our list that we are sure is unique is the Social Security Number. With the Social Security Number as the primary key, our table is in the second normal form. When we decided to not put the pay rate in the table, we put the table in the third normal form (again, because we make every value depend only on the primary key, in this case the Social Security Number). Here is the table.

Check This Out...

Why Two Programmers Are OK Looking at the fact that there are two rows that contain identical values (programmer) in the following table, we see again how we can begin to decide on the primary key. Two separate unique people can have jobs with the same title. They are still unique persons even though they have some characteristics that are the same. The Social Security Number is always unique. No two people will share the characteristic of the same Social Security Number.

People Table

SSN	First Name	Last Name	DOB	Job Title
111-22-3333	Sam	Jones	05-25-1955	Programmer
222-33-4444	Tom	Smith	07-14-1964	Secretary
333-44-5555	Dorothy	Rubyslippers	08-29-1981	Programmer

Phone Number Table

The next table will be the Phone Number Table. Since the purpose of the table is to show which person has which phone number or numbers, we need three columns:

- Area Code
- Phone Number
- Social Security Number

SSN not Name The reason that we chose Social Security Number rather than name is that it is the primary key in the People Table. It is the primary key in that table because it is the one property that will always be unique.

This table is in the first normal form since all values are atomic. What can we select as the primary key? Since a person can have more than one phone number and a phone number can be used by more than one person, and further, the same phone number will appear in multiple area codes, our only choice is a combination of all three columns. This will always be unique.

However, we do have an alternative. We can add a unique number to each record and use that number as the primary key. When you look at the table below, you see that is what we have done.

Phone Table

Key	SSN	Area Code	Phone Number
1	111-22-3333	202	555-1212
2	222-33-4444	212	555-1345
3	111-22-3333	202	444-9876
5	333-44-5555	316	444-4321
7	222-33-4444	212	555-1345

The Foreign Key

When you examine this table, you see that we don't have a name, just the SSN from the other table. The SSN is called a *foreign key*. When we want the name that goes with a phone number, we will get information from two tables. To connect the two tables we need an item of data that appears in both tables. This item of data is the primary key in one table and a foreign key in the other table.

The Job Table

The next and last table that we need to create is one for the job title and pay rates. This will be the Job Table.

Job Table

Job Title	Pay Rate
Programmer	$25.00
Secretary	$15.00

The benefits of RDBMS are easy to see. For example, picture a company of thousands of employees and 30 job titles. If we had the pay rate change for all of the job titles, say a cost of living raise, we need change only the 30 job title records and everyone's pay rate has been changed. Each item of data is recorded only once, so if it needs to be changed, there is only one place that needs to be changed. As an example, if Dorothy Rubyslippers married Fred Combatboots and decided to change her last name to that of her husband, there would be only one change needed to have Dorothy's name change recorded throughout the database and any application systems that use the database.

Structured Query Language—SQL

SQL is used to retrieve data from the relational database, to add data to the database, to delete data from the database, and to change data in the database. The nice thing about SQL is that it is similar to ordinary English. (SQL can be a complex and frustrating topic to learn. Here we are only introducing SQL.) In Chapter 13, "Query Designer, or Instant SQL," you will see that Visual InterDev provides a tool that will relieve you of the burden of learning SQL in exquisite detail.

Select Statement

If we wanted to retrieve the last names from the People Table, the SQL statement would be:

SELECT Last Name FROM People

The syntax of the statement is as follows:

- **SELECT** Tells what we are going to do.
- **Last Name** Is the name of the column from which we are retrieving data.
- **FROM** This key word tells that what follows is the name of a table.
- **People** This is the name of the table from which data will be retrieved.

The database would return a set of columns and rows, called a results set, that would look like this:

People
Jones
Smith
Rubyslippers

Joins

If we want to retrieve a phone list from the tables, the SQL Statement would be:

SELECT First Name, Last Name, Area Code, Phone Number FROM People, Phone WHERE People.SSN = Phone.SSN.

The syntax of the statement is as follows:

- **SELECT** Tells what we are going to do.
- **First Name, Last Name, Area Code, Phone Number** This is the column name list. It is comma separated.
- **FROM** This key word announces that what follows is a list of tables.
- **People, Phone** This is the list of table names separated by commas.
- **WHERE** This key word announces that we are doing a logical join of two tables.
- **People.SSN** This is a qualified column name. It is the SSN column from the People Table.
- **=** This is the logical operator stating that both sides must be an exact match.
- **Phone.SSN** This is another qualified column name. It is the SSN column from the Phone Table.

The results set will look like this:

First Name	Last Name	Area Code	Phone Number
Sam	Jones	202	555-1212
Tom	Smith	212	555-1345
Sam	Jones	202	444-9876
Dorothy	Rubyslippers	316	444-4321
Tom	Smith	212	555-1345

The columns are in the order from left to right that was specified in the SELECT statement. SSN does not appear even though it was used as a criteria. This type of select is called a JOIN because the data from two tables are joined in one results set.

The results set can be sorted by adding an ORDER BY clause to the select statement. The SELECT statement would then read:

SELECT First Name, Last Name, Area Code, Phone Number FROM People, Phone WHERE People.SSN = Phone.SSN ORDER BY Last Name.

Views

A view is a very useful tool in the world of relational databases. A view is a combination of data that looks like a table, acts like a table, and that the end user can't tell from a table, but it exists only as a SELECT statement-like definition. If you have ever worked with Microsoft Access, you know that an Access Query is a view. A view is defined within the database; for your purposes, you can also treat a view as a table.

Update Statements

When we wanted to change Dorothy's last name from Rubyslippers to Combatboots we could have used an UPDATE statement:

UPDATE People SET Last Name = 'Combatboots' WHERE Last Name = 'Rubyslippers'

There may be a problem with this. The RDBMS (also called the database engine) finds each instance of Rubyslippers in the Last Name column of the table People and changes it to Combatboots. This can be a problem if there are others with the Last Name of Rubyslippers, so we should use the Social Security Number in the WHERE clause. The primary key (SSN) is unique, so only one record will be changed:

UPDATE People SET Last Name = 'Combatboots' WHERE SSN = '333-44-5555'

The syntax of the statement is as follows:

- **UPDATE** This tells what we are going to do.
- **People** This is the name of the table being updated.
- **SET** This key word announces that what follows is a column name and the value that will be placed in the column.
- **Last Name = 'Combatboots'** The name of the column being updated and the value being placed in the column.
- **WHERE** Announces that what follows is a logical selector.
- **SSN** The name of the column containing the value being tested.

- **=** The match must be exact.
- **'333-44-5555'** The value that must be found. This should usually be the primary key.

Don't Forget the Where Clause

This is a much safer approach. A point to think about is what would be the result if we used the statement with no WHERE clause as this:

UPDATE People SET Last Name = 'Combatboots'

If you guessed that every record in the database would be changed, you are correct.

Delete Statement

The delete statement is a very simple—and very dangerous—statement. If we wanted to delete the record for Dorothy from the People Table, the statement should be:

DELETE FROM People WHERE SSN = '333-44-5555'

The syntax of the statement is as follows:

- **DELETE** This tells what we are going to do.
- **FROM** This key word announces that the following is a table name.
- **People** The name of the table that is going to be changed.
- **WHERE** Announces that the item that follows is a logical selector.
- **SSN** The name of the column containing the data being selected.
- **=** The match must be exact.
- **'333-44-5555'** The value being matched.

In this case only one record is affected. When it is removed from the table, it is gone. There is no UNDO function on most databases. That is why a confirm window frequently is used to be sure that you really, really, really want to delete the record.

Can you predict the result of the following SQL statement:

DELETE FROM People

Right you are. Every row in the table is gone.

Insert Statement

Records are added to a table using an INSERT statement. For example, let's insert a row. One of the tasks that the database engine will perform is to check that the primary key that we have used for the new record is unique. If it is a duplicate, the INSERT operation will fail and no change will be made. An example of an Insert statement might be the following that was used when Tilly Tennisshoes was hired:

INSERT INTO People (SSN, First Name, Last Name, DOB, Job Title) VALUES ('999-88-7777', 'Tilly', 'Tennisshoes', '12-18-1959', 'Programmer')

The syntax of the statement is as follows:

- **INSERT INTO** This tells what we are going to do.
- **People** This is the name of the table into which we are inserting a record.
- **(column list)** This is the list of columns in the table into which we will be inserting data. The list is a group of items separated by commas and enclosed by parentheses.
- **VALUES** This key word tells that what follows are the values to be inserted into the columns.
- **(values list)** This is the list of values that are separated by commas and enclosed in parentheses.

The Least You Need to Know

This has been a very brief, very shallow dip into the waters of the world of RDBMS. There are volumes written on the subject of SQL alone; if you're really interested—or in need of a good SQL tome—I suggest you check out *Special Edition Using MS SQL Server 6.5*, published by QUE. Creating SQL statements is somewhat like chess. Knowing the moves does not make you a master. However, the goal was not to turn you into an SQL master; but rather to make it possible for you to read an SQL statement with a degree of understanding. Just remember:

- Data is contained in tables.
- Tables should have a primary key.
- Data from multiple tables can be combined through a JOIN.
- Data is retrieved using SELECT statements.
- INSERT statements are used to add rows to a table.
- WHERE clauses are used to select the rows that will be affected.

- UPDATE statements should have a WHERE specifying the primary key.
- DELETE statements may be dangerous. A WHERE clause for the primary key will have the effect of deleting only one row.
- There is no UNDO for a DELETE or UPDATE. It may be wise to have a confirmation by the user before the operation is completed.

Chapter 13

Query Designer, or Instant SQL

In This Chapter

- Establish an ODBC Data Source Name
- Connect to an ODBC Data Source
- Use the graphical features of the Query Designer to create SQL Statements
- Save query files for future use
- Review a results set to see if the query produces the desired results
- Verify the query that you have created

Wouldn't it be nice if you had an assistant that worked for almost nothing (just electrons) and did all of the SQL creation that you needed for your applications?

Well look no further—just such an assistant has arrived in the form of the Visual InterDev Query Designer! This query designer is guaranteed to produce flawless, syntactically correct SQL statements each and every time. (That is if you know how to ask it the proper question so it will know exactly what you want.)

Types of Queries

There are four types of queries that you will want to work with in your database manipulations. These are:

What is a Query?
Query, as it is used here, is another name for an SQL Statement. SQL was discussed in Chapter 12 "Relational Databases and SQL: Information, Please!" If you are having trouble with this discussion, you might want to look over Chapter 12.

- **Select Query** This query is used to retrieve data from a database. An example is—SELECT FirstName, LastName FROM NameList. In this example, FirstName and LastName are column names in the database table NameList.
- **Update Query** This query is used to change data in an existing database record. An example is—Update NameList Set FirstName = 'Sam' Where LastName = 'Smith'.
- **Insert Query** This query is used to add a new record to a table. An example is—Insert INTO NameList (FirstName, LastName) Values (Fred, Peters).
- **Delete Query** This query is used to remove a record from a table. An example is—Delete From NameList Where LastName = 'Smith'.

In this chapter, we will be using the Query Designer to generate SQL statements. We won't be using the queries in Web pages here, that comes later in Chapter 15, "Database Connectivity—Long Distance, Please." It is enough now to be able to create these four types of SQL queries with the Query Designer.

Connecting to a Database

To begin, we must establish a connection to a database using Open Database Connectivity (ODBC). For this example, we are going to connect to an SQL Server database named "pubs". (If you have worked with SQL Server, you will be familiar with pubs; it is a sample database that installs with SQL Server for use in testing.)

Check This Out...

No SQL Server? If you don't have access to an SQL Server, these instructions may be followed with some slight alterations for any other database for which you have an ODBC driver installed on your system. If you have Access on your system, you probably will have a "North Wind" sample database available, which will also work just fine.

First, we need to open Visual InterDev and choose **File**, **New** on the menu bar. When the New dialog box opens, click the **Projects** tab and highlight the **Database Project** selection. Now enter a name in the Project Name text box. The name shown in the following figure is "CIG_Qdesign".

After entering the project name, click the **OK** button. You are now presented a dialog box titled Select Data Source, as shown in the following figure. This is the beginning of the ODBC setup.

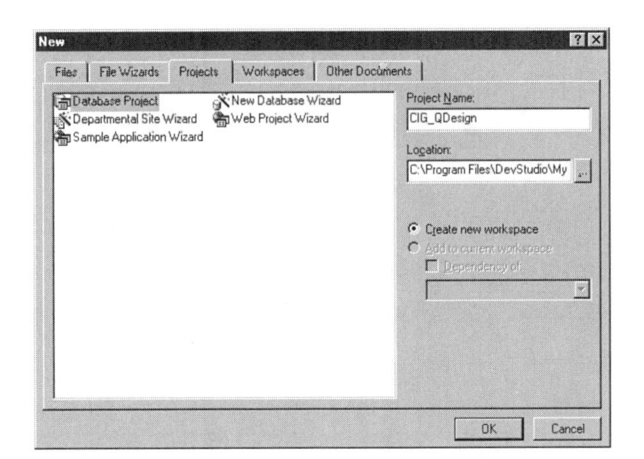

The project name should be a descriptive name that has meaning to you.

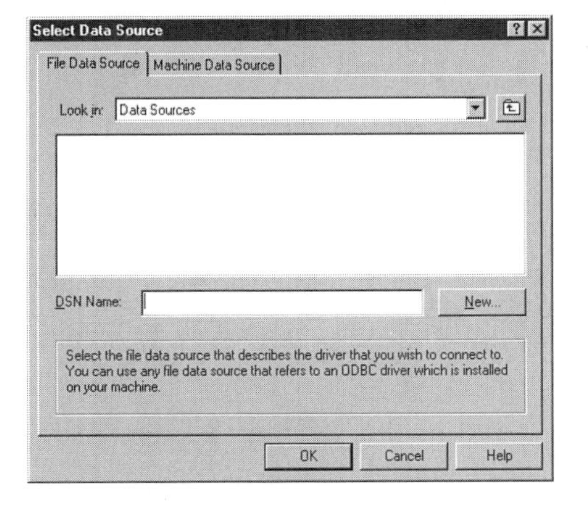

This is the ODBC Administrator that has been opened by the Database Project Wizard of Visual InterDev.

Click the **Machine Data Source** tab and you are presented with the list of data sources. Since none of these are the pubs database, we need to click the **New** button.

You are now looking at the Create New Data Source dialog box, as seen in the next figure.

Select the **System Data Source** radio button and click the **Next** button at the bottom of the dialog box. The next dialog box, also labeled **Create New Data Source**, is used to choose the ODBC driver for the database to which you will be connecting. Highlight **SQL Server** as shown in the following figure, and click the **Next** button. You will be presented with a confirming dialog box that recaps what is about to happen—creating a System Data Source using the SQL Server ODBC driver. Click the **Finish** button.

All of the data sources that you set up for use with ODBC and Web pages must be System Data Sources.

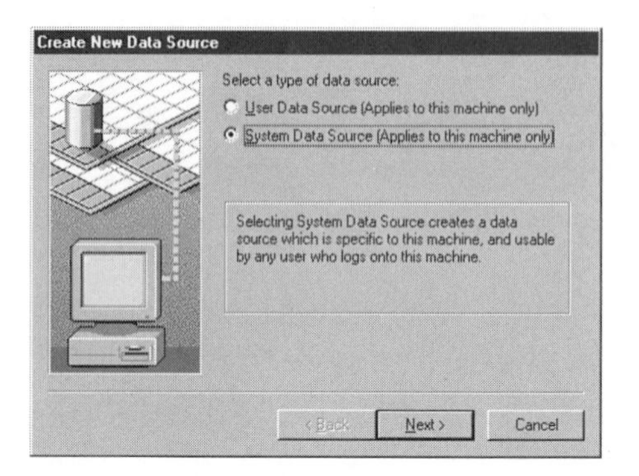

There are ODBC drivers for almost all of the relational databases in use today. Only a few are shown here.

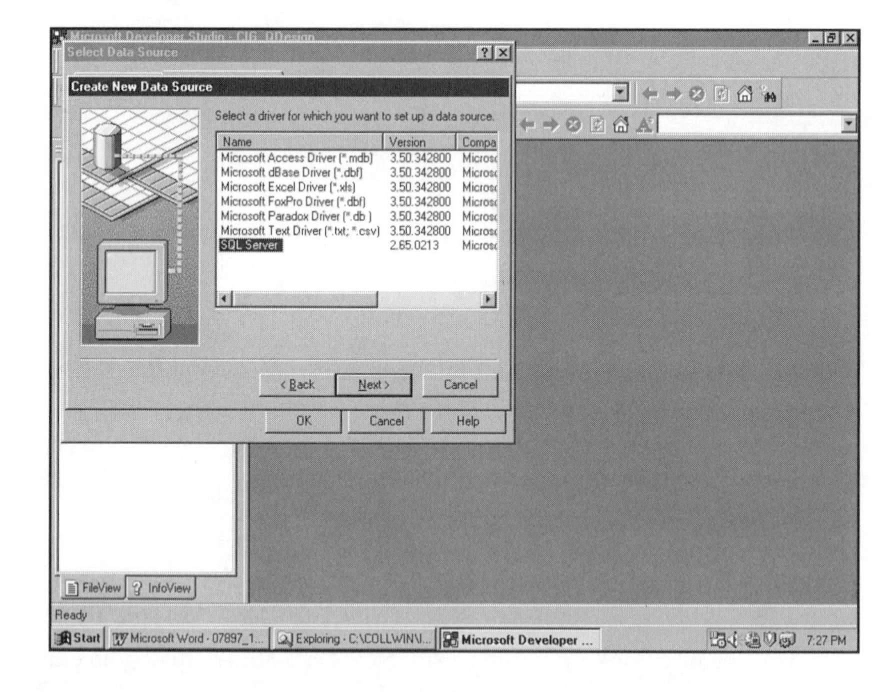

System DSN

A System Data Source is one that can be seen by multiple programs on the computer. For all work with Web pages and Active Server using databases, ODBC data sources must be System Data Sources.

You are now presented with a dialog box labeled ODBC SQL Setup. The first thing that you need to do is click the **Options** button on the lower-right of the dialog box to open the remaining part of the dialog box that we will be using. Enter the Data Source Name in the text box. (CIG_QDesignDSN has been used in the next figure.) Next, enter the name of the server in the Server text box. You may want to check the Use Trusted Connections check box and enter **pubs** in the **Database Name** text box. (If you have questions about what some of these entries are, see your database administrator for assistance.)

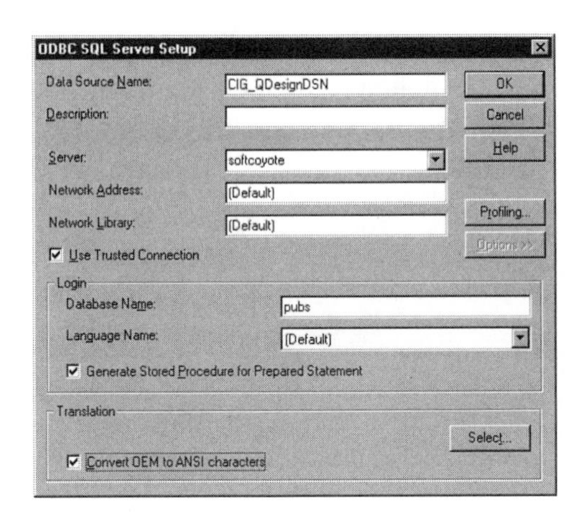

The description is always an optional entry.

Now click the **OK** button. You have just created an ODBC connection.

Designing a Select Query

The Workspace window (usually on the lower-left of your screen) has changed to reflect two additional tabs on the bottom. These tabs are the File View and Data tabs, in addition to the Info View tab. Click the **Data** tab and click the + symbols to the left of the objects to open up the data list as shown in the following figure .

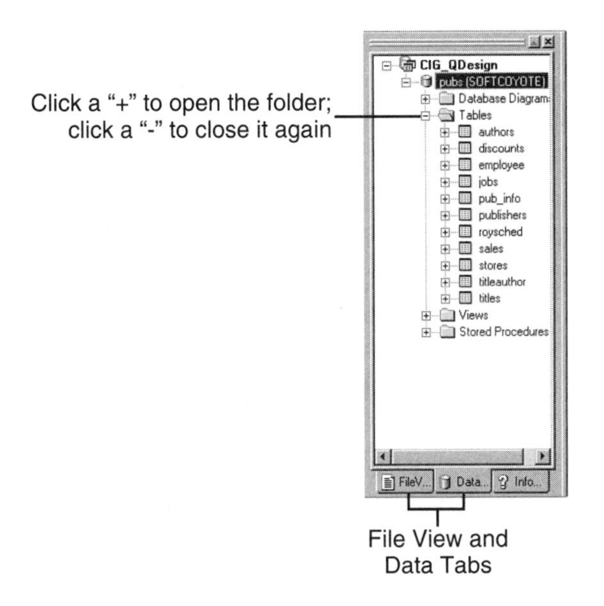

Double-clicking one of the tables will open the table and display the data.

You are now ready to begin to design the select query. On the menu bar choose **Insert, New Database Item** and the dialog box shown in the next figure titled Insert Database Item will appear. Highlight **Query** and click the **OK** button.

This same dialog box is used to insert a number of different new database objects.

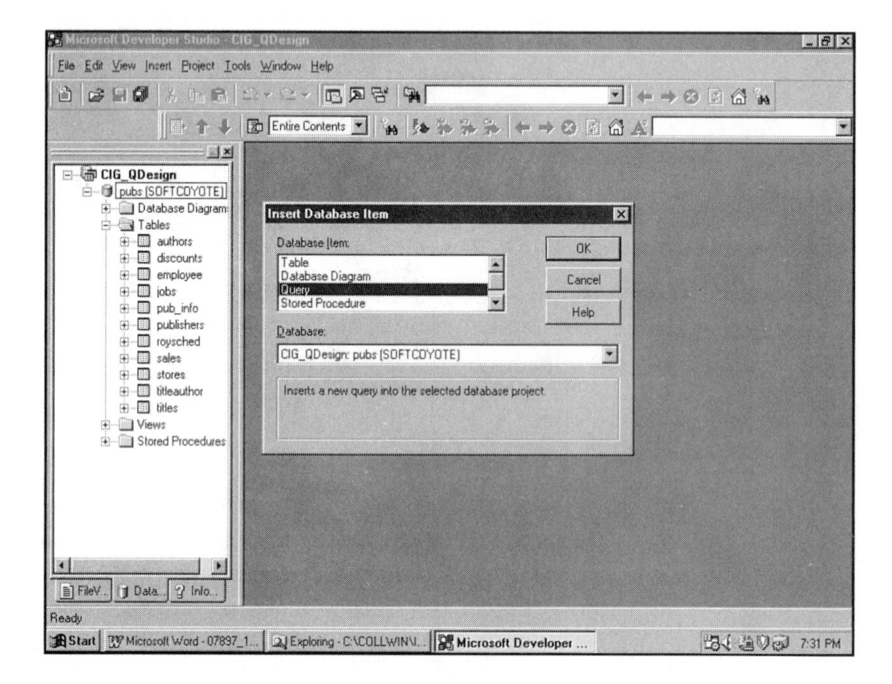

You are now presented with the Query Designer dialog box. This dialog box is where we will be performing the design work. Let's pause and examine the parts of this so we will understand that with which we will be working.

- **Query Tool Bar** When the Query Design dialog box opens, this tool bar is automatically opened. It contains a number of useful tools.
- **Diagram Pane** This is the pane in the Query Design dialog box where the tables that are being used in the query are shown.
- **Grid Pane** This is the pane where the columns that are used in the query are shown.
- **SQL Pane** The SQL statement that is created by the Query Designer is displayed here.
- **Results Pane** After a query has been designed, the query can be run to see if it produces the correct results.

- **Show Diagram Pane Icon** This icon is a toggle to turn on the display of the Diagram Pane.

➤ **Show Grid Pane Icon** Toggle for the Grid Pane.

➤ **Show SQL Pane Icon** Toggle for the SQL Pane.

➤ **Show Results Pane Icon** Toggle for the Results Pane.

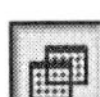

➤ **Create Select Query Tool** Click this tool to create a Select query.

➤ **Create Insert Query Tool** Click this tool to create an Insert query.

➤ **Create Update Query Tool** Click this tool to create an Update query.

➤ **Create Delete Query Tool** Click this tool to create a Delete query.

➤ **Verify SQL Tool** Click this tool after an SQL statement has been designed and the statement will be checked against the table to see if it is correct and will work.

➤ **Run Query Tool** Click this tool after an SQL statement has been designed and is run with the results being shown in the Results Panel. Warning, Delete, and Update queries make the changes to the database with no undo available.

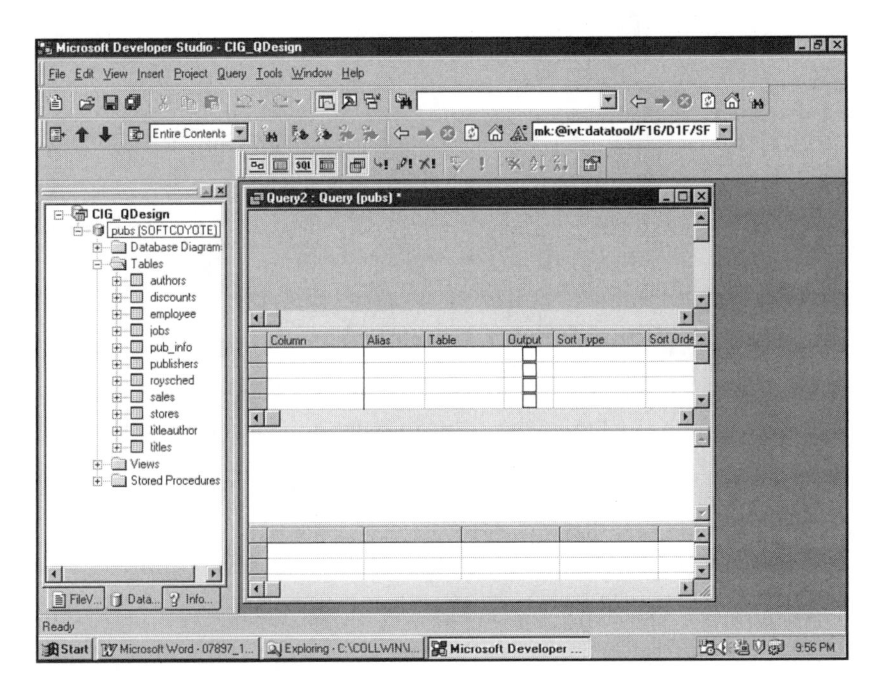

There are four panes in this dialog box, each of which we will be using in the design process.

Designing the Select Query

Now we will design the Select Query. For the demonstration, we are going to select author ID, first name, last name, and phone number from the authors table where the last names start with the letter D or after in the alphabet. We then want this list to be sorted by last name and first name. (Don't ask why we want this list. It just sounded like a challenge to create.)

To include the authors table in our query, place your mouse over the authors table in the Workspace window on the left, left-click and hold while you drag the table to the Diagram Pane of the Query Design dialog box. Now release and you will see the table in the Diagram pane with the list of columns displayed for selection. Clicking the check box to the left of the column name will add the column to the query as shown in the following figure.

*The *(All Columns) can be checked to add all columns to the query. This saves time if you are going to use all of the columns.*

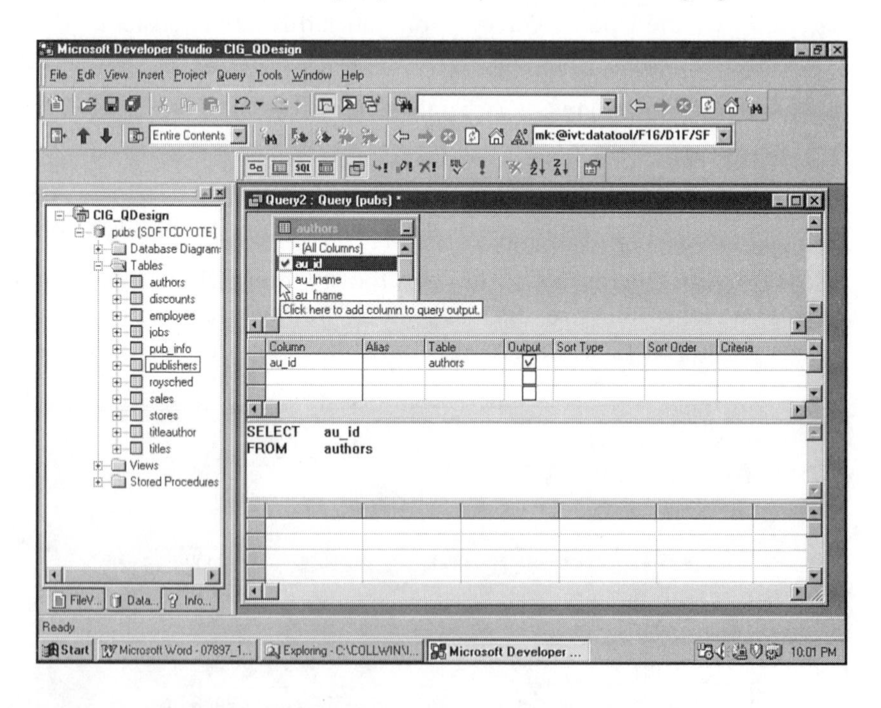

Valid SQL
Did you notice that SQL begins to appear in the SQL window? If you checked this SQL statement with the Verify SQL tool, you would find that it is a valid statement.

Now it is time to add the other three columns that are **au_lname**, **au_fname**, and **phone** by checking the check box to the left of the column name in the Diagram Pane. After these columns are added, we need to establish the sort order. Take your mouse and left-click the **Sort Type** column of the Grid Pane next to the au_lname column. The words "Ascending" and "Descending" will appear in a drop-down list. Select **Ascending**. Now repeat the procedure for the au_fname column.

Two things have now happened. First, the Sort Order column of the Grid Pane now has numbers. These numbers indicate that the results will be sorted by last name and then first name. Second, an ORDER BY clause has been added to the SQL statement in the SQL Pane, as seen in the following figure.

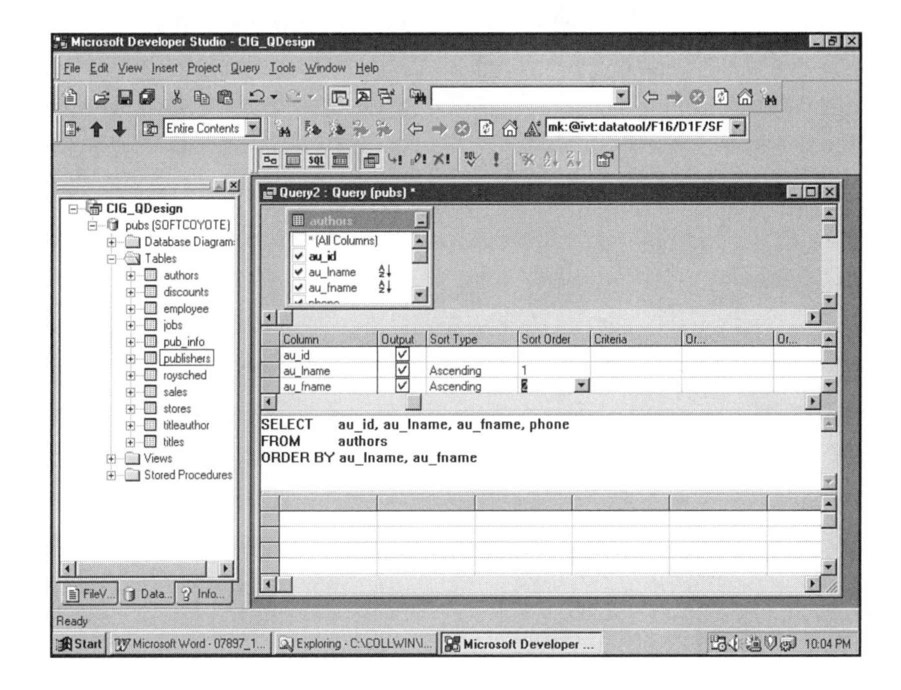

The SQL statement in the SQL Pane reflects the graphic changes that have been made in the Grid Pane.

We are now ready to add the criteria that will select only records with last names starting with the letter D or further in the alphabet. After the criteria has been added, we will run the query, look at the results set, and if it looks proper, we will save the query in a file that can be reopened later.

To set the selection criteria for the letter D, place the mouse in the Criteria column in the row for last name in the Grid Pane. Click to set the focus to this cell. Now enter the greater than symbol > and type **'D'**.

Now it is time to run the query. Click the **Run Query** tool on the Query tool bar. The result should look like the following figure.

If everything looks proper, we are ready to save the query in a file. Choose **File**, **Save As** in the menu. Browse to the directory that contains the project, in this case C:\Program Files\DevStudio\MyProjects|CIG_Qdesign. Set the file name as CIG_Qdesign_Select.DTQ and click **OK**.

You have successfully completed the design of your first Select Query using the Query Designer! Now, close the Query Designer and you will be ready to design the next query, an Update Query.

The last names in the results set should begin with the letter D or further and be sorted into alphabetic order.

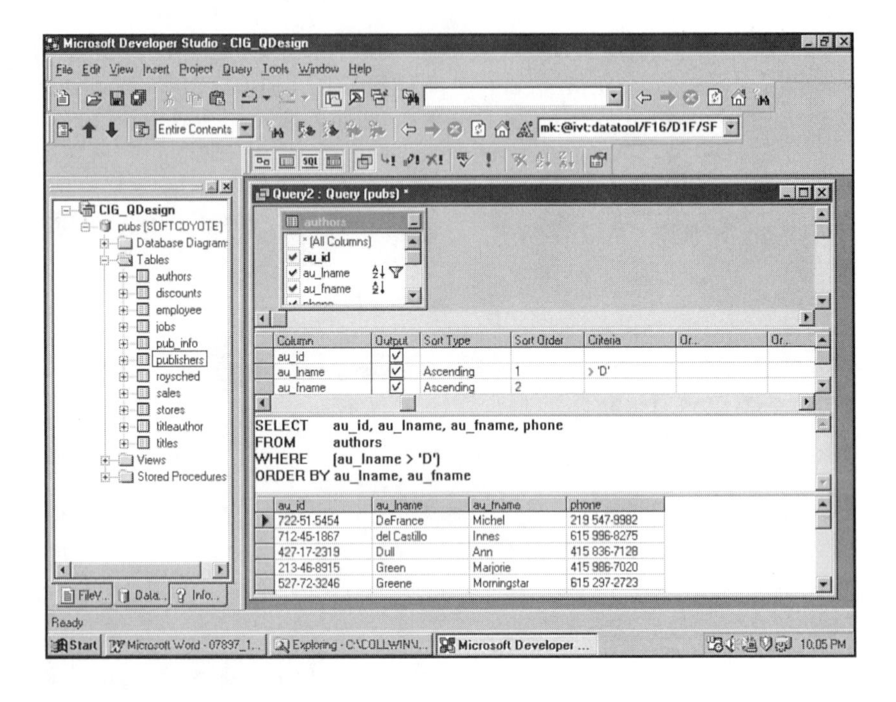

Designing the Update Query

1. With Visual InterDev open and the CIG_Qdesign Workspace open, choose **Insert**, **New Database Item** to open the Insert Database Item dialog box.

2. Highlight the Query selection and click **OK**.

3. Now click the **Create Update Query** icon on the Query toolbar.

4. Next drag the authors table to the Diagram Pane. Choose the **au_id**, **au_lanme**, **au_fname**, and **phone columns**. Now enter the values that are shown in the New Value column of the Grid Pane. These values are au_lname—'Smithers', au_fname—'Georgia', and phone '508-322-7755.

5. Enter the following in the Criteria column next to au_id—= '172-32-1176'. Your design dialog box should now look like the figure below.

We are now ready to run the query. Click the **Run Query** icon on the Query toolbar. If the Query runs successfully, you will see a message box as shown in the following figure announcing that one row was affected.

Finally, save the file as before with a name of CIG_Qdesign_Update.DTQ. and close the Query Designer.

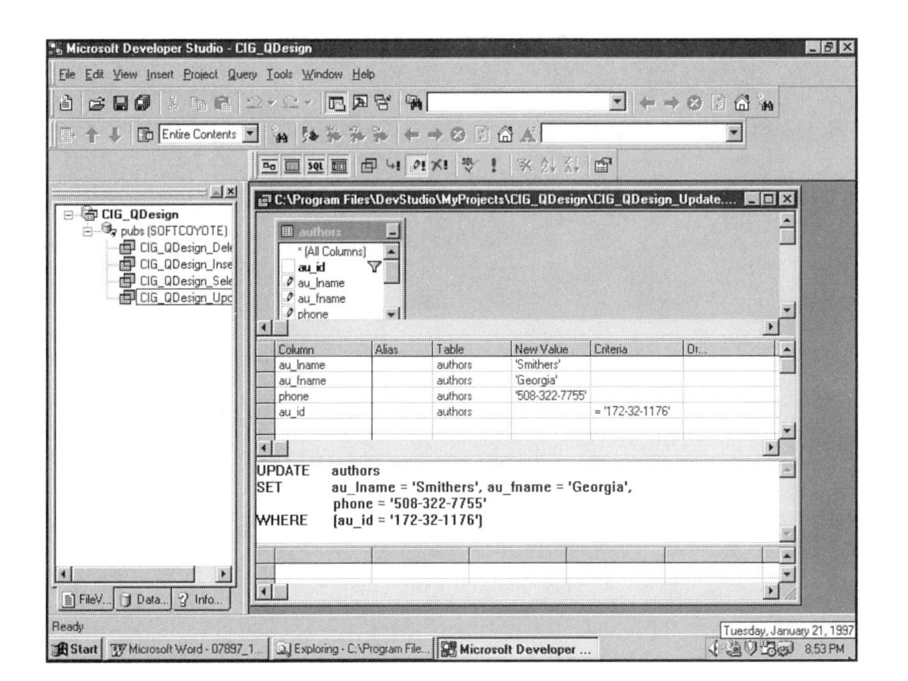

You can test the SQL statement before running the Query by clicking the Verify SQL tool in the toolbar.

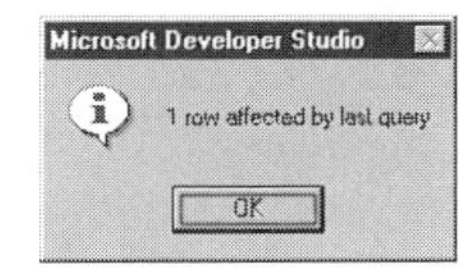

The Message Box announces the results of the query. In this case there is not a results set that is returned by the query.

Designing the Insert Query

By now you are an old hand at opening the Query Designer. This time you will select the Create Insert Query type. (In preparation for this exercise, an empty table was added to the pubs database named Table1. The table contains four columns: au_id, au_lname, au_fname, and phone with the identical characteristics as the authors table.)

When you click the **Create Insert Query** icon a dialog box opens asking you into which table you are going to insert records. Drop the list of tables and select **Table1**. Now drag the authors table to the Diagram pane. Select the columns **au_id**, **au_lname**, **au_fname**, and **phone**. The Query Design window should now look like the next figure.

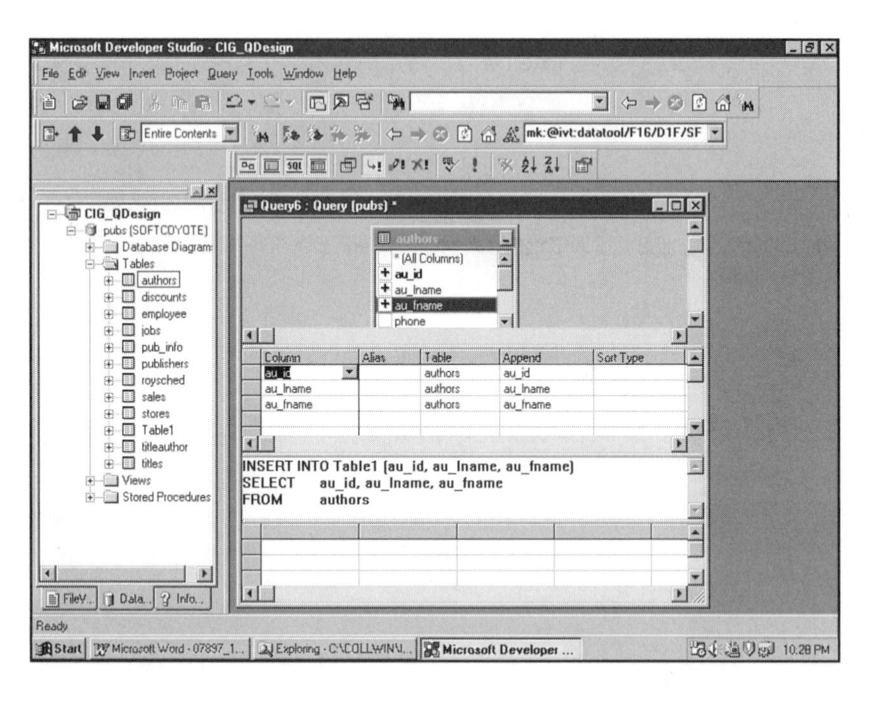

This will copy data into a new table from an existing one.

Test the SQL statements using the **Verify SQL** tool on the Query toolbar. If it checks out, run the query by clicking the **Run Query** icon on the Query toolbar. You will see a Message Box, as in the following figure, that announces that 23 rows were affected. In other words, 23 rows were copied from one table to the other. Now save the query in a file named CIG_Qdesign_Insert.DTQ.

Finally, close the Query Designer in preparation for creation of the last query.

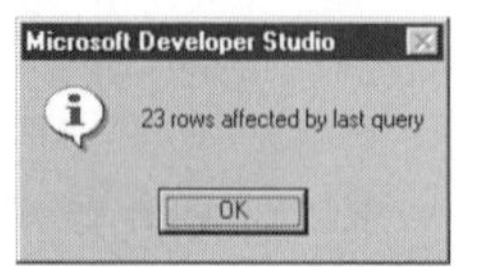

This query does not return a results set. It instead announces the number of rows that have changed as a result of its operation.

Designing the Delete Query

One more query to go and you will have covered the four major query types.

1. Take heart—you are on the home stretch. Open the Query Designer and click the **Create Delete Query** icon on the Query toolbar.
2. Drag Table1 to the Diagram pane. In the Grid Pane, click the top cell in the "Column" column and a drop-down list of the column names appears.

3. Select **au_lname** and any other columns that you want to set criteria for.
4. In the Criteria column type > **'M'**. This will delete all records with a last name starting with M or after in the alphabet. The Query Design dialog box should look something like the figure below.

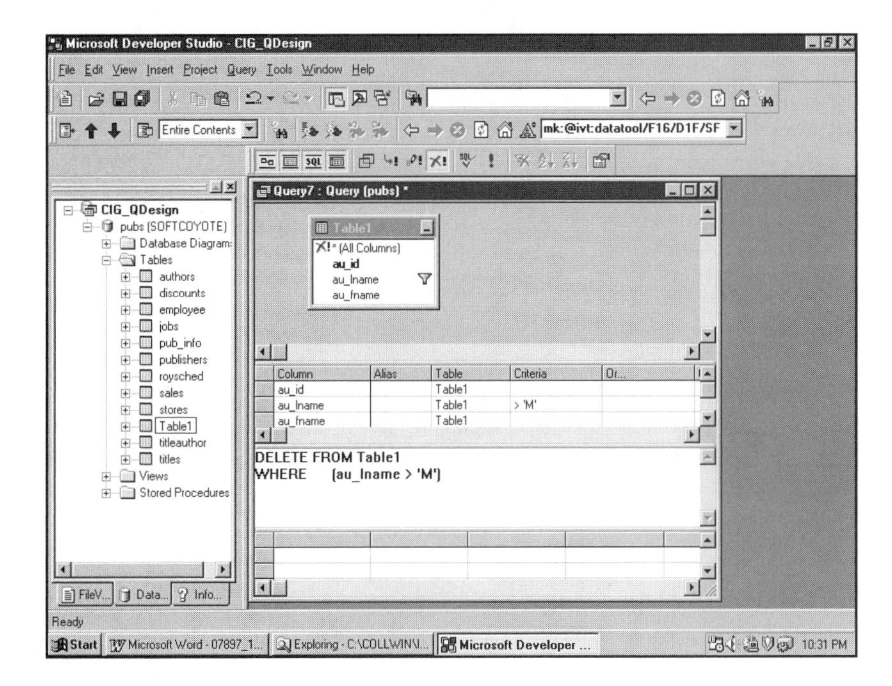

When a Delete Query is run, there is no undo option. The data is deleted forever. Make sure that you want to do it before you click run.

Now test the SQL statement with the Verify SQL tool. If it checks out, click the **Run Query** icon on the Query toolbar. You will see the now familiar announce as shown in the following figure, that 11 rows were affected.

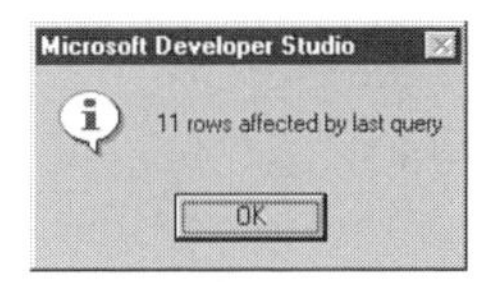

The Delete query has no results set to return except the announcement that 11 rows were deleted.

Save the query in a file named CIG_Qdesign_Delete.DTQ. Pat yourself on the back. You are now an accomplished query designer.

If you click the File View on the Workspace Window, you will see the four query files that have been created. If you click the CIG_Qdesign_Select, you will see the Query Designer open, (with the query that you created) all in place for review or modification as shown in the next figure.

Saving your queries as files allows you to reuse queries that are similar to what you now need.

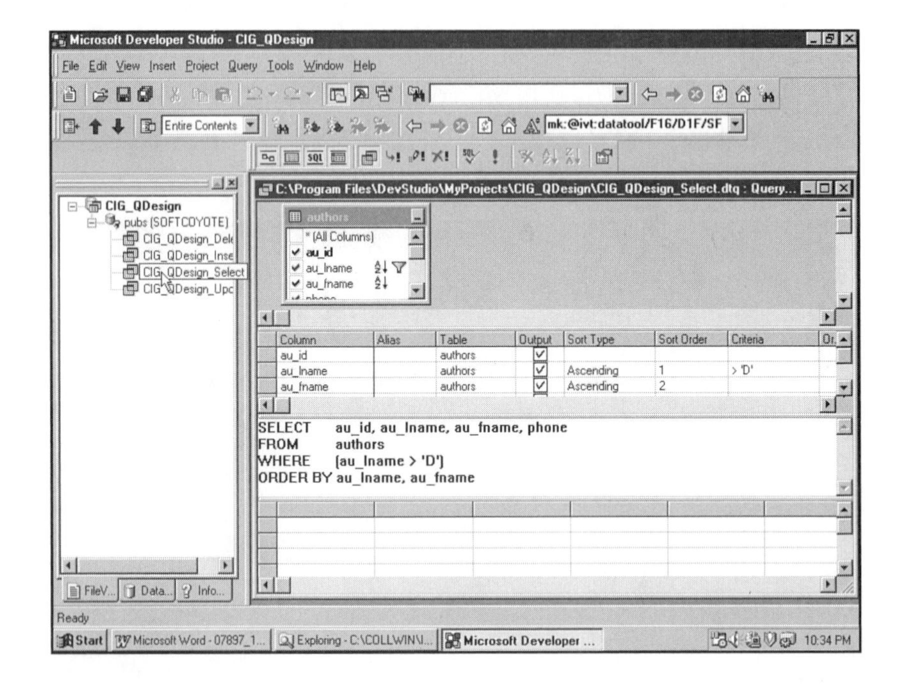

The Least You Need to Know

As you work further with the Query Designer, you will see that we have only scratched the surface of this powerful, excellent tool. For example, one feature that we haven't mentioned is that you can copy the SQL generated in the past into an HTML form when you need a SQL statement (with correct and testable syntax). You will be able to avoid the hours attempting to get an SQL statement to do your bidding.

- ➤ Writing SQL can be fraught with hazard. Letting the Query Designer create your syntactically correct SQL statements will save you many hours and much frustration.
- ➤ Query Designer SQL can be used other places in a Web site. Any time you need an SQL statement, turn to the Query Designer for assistance.
- ➤ Learning SQL requires many long hours of concentrated study. Learning the Query Designer is a quick and fun exercise.

Chapter 14

Database Wizards, or Instant Databases

In This Chapter

- Create a new SQL Server database
- Create a database project with an existing database

Visual InterDev provides two database Wizards that are very useful to both the experienced database user and the database novice. For the novice, these Wizards will make databases usable without requiring extensive training. For the experienced user, the Wizards save time by creating a new database and database projects.

Creating a New Database

The creation of a new database is possible only with the Microsoft SQL Server Version 6.5 Relational Database Management System (RDBMS). The best way to understand how this Wizard works is to use it.

SQL Server Requirements

In order to use SQL Server with Visual InterDev, you must have SQL Server 6.5 with SQL Server Service Pack 2 installed. The service pack is in the \Server\Sqlsrvsp directory on the Visual InterDev install CD. The setup program is named Setup.exe. You should be aware that you will need to stop SQL Server to install the service pack but you will not be required to reboot the server. The installation of the service pack will require a minimum of about 20 minutes.

Using the New Database Wizard

To create a new database: with Visual InterDev open, choose **File**, **New** and click the **Projects** tab. Highlight the **New Database Wizard** selection and enter a name for the project in the **Project name:** text box, as shown in the figure below.

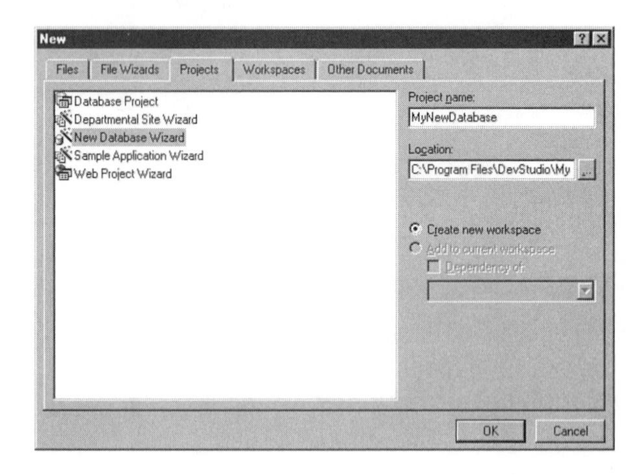

The Create new workspace radio button is selected. This is the only option since there wasn't a workspace open when the process was started.

Click **OK**. You are now presented with Step 1 of 4 of the New Database Wizard. You will need to enter three items of information in this dialog box. After you have entered the data, click the **Next** button.

- **SQL server** This is the name of the SQL server on which you are going to create the database.
- **Login ID** This is a login that is setup on the SQL server with Create Database privileges. In the example, the Login ID used is "sa".
- **Password** This is the SQL server password for the Login ID. In the example, the password used is the password for the sa.

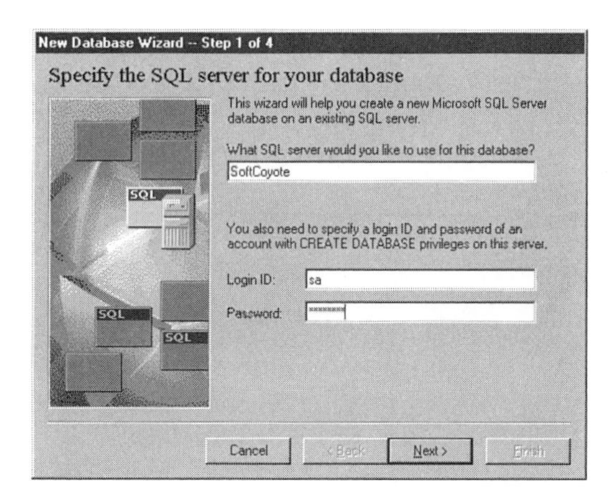

If you do not have Create New Database privileges on the SQL server you select, contact the Database Administrator for assistance.

Step 2 of 4 of the New Database Wizard requires the selection of the SQL Server database device upon which the database is to be created. You can select to have the new database created on an existing device or a new database device.

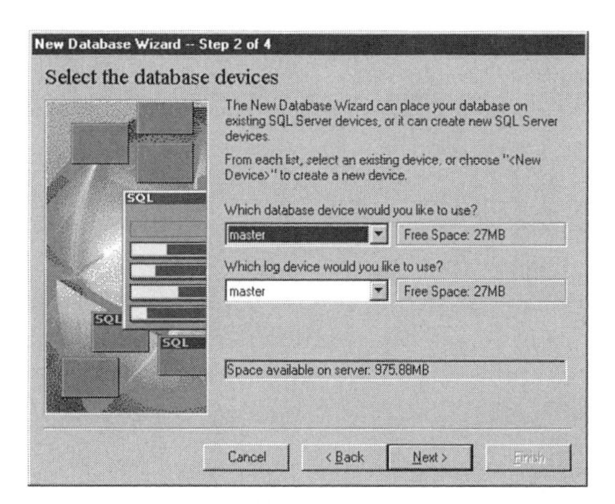

The dialog box displays the available space in the existing database devices and available space on the server.

SQL Server Database Devices

In SQL Server, the database device is the physical file on the server in which a database is created. Before a database can be created, the database device must exist. A database device file has a .DAT extension. When Visual InterDev is used to create a new database device, it is created on the server in the directory \MSSQL\Data.

We are going to create a new database device. Start the process by clicking the drop-down list arrow on the database device list box and selecting **<New Database Device>**. This opens the New Device Information dialog box as shown in the figure below. In this dialog box we need to supply the device name and the device size in megabytes.

The minimum device size is 1 megabyte.

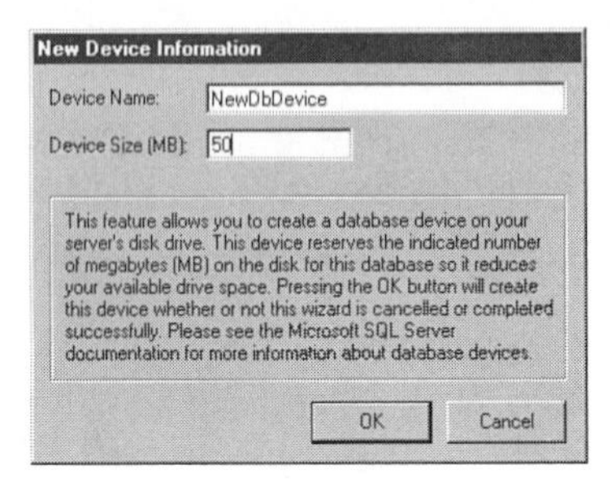

After creating the new database device, you are returned to the Step 2 of 4 dialog box. It is now time to select a log device for the database. A separate device may be chosen for the log and a new database device may be created for the log. After you select the log device, click the **Next** button.

Techno Talk

SQL Server Logging

The log for a SQL Server database is used in two circumstances. The first and most common is for transaction processing. Briefly, a database transaction is a series of database changes that all must be correct and accepted before you want to make any changes. In transaction processing, if any step of the change fails, the entire transaction is rolled back to its condition before the transaction began. This type of action requires a "before and after picture" to be taken of the affected portions of database. These pictures are stored in the database log.

The second circumstance that the logs may be used is in the event of a power failure or system crash that occurs in the middle of an update, for example. The log is used to rollback when SQL Server is restarted.

The Step 3 of 4 dialog box of the New Database Wizard asks for three pieces of information as shown in the next figure. After you enter the information, click the **Next** button.

- **Database name** The default name supplied is NewDatabase. This can be any name that fits the SQL Server naming rules. We used the default in this instance.
- **Database size in megabytes** The minimum is 1 MB and the maximum is the size of the available space on the database device. There are complex formulas for the calculation of the size of an SQL Server database. You can always increase the size of the database if need be.

➤ **Log size in megabytes** The size parameters are the same as the database. The defaults shown in the example are 5 MB for the database and 2 MB for the log. This ratio of 40% for the log is not based on any magic formula. There are formulas that will help you calculate the correct size for the log.

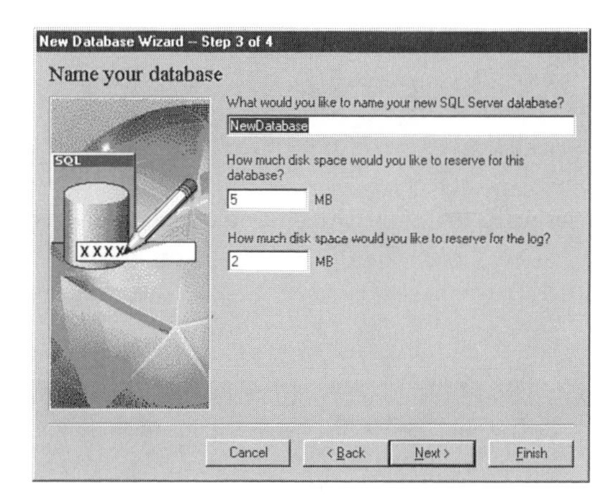

If you have made any error, you can click the Back button to return to the prior dialog box to correct the error.

Step 4 of 4 of the New Database Wizard requires no information. Unless you have made an error that must be corrected (click **Back** to make corrections), click the **Finish** button (if you click **Cancel**, the Wizard is closed and all information entered is lost).

Creating Tables in the New Database

You now have an SQL Server database. It is an empty database with no tables or data. If you click the expand tree symbol (+) on the database objects in the Workspace Window, you will find that there is nothing there.

Your next task is to create a table in the database. Choose **Insert**, **New Database Item** and highlight the selection Table as shown in the figure below.

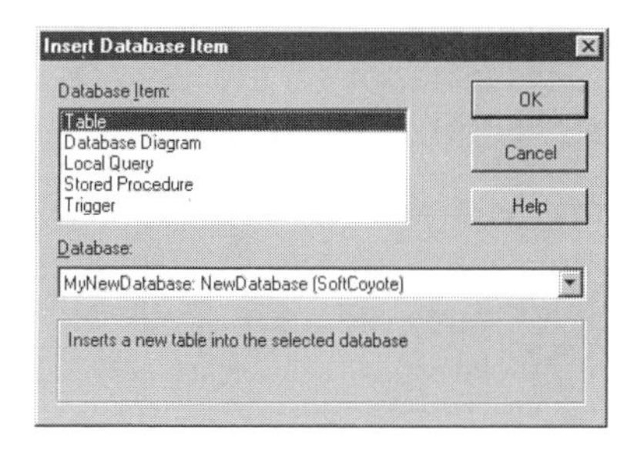

Notice that the project name (My New Database), the database name (New Database), and server name (Soft Coyote) are shown in the list box. Another database in the project may be selected.

Click **OK**. The next dialog box is the Choose Name dialog box. The name for the table First_Table is the name I gave to the table as shown in the figure below (making fun of the names I choose is not allowed).

If all dialog boxes were this self explanatory, authors like me would be out of a job.

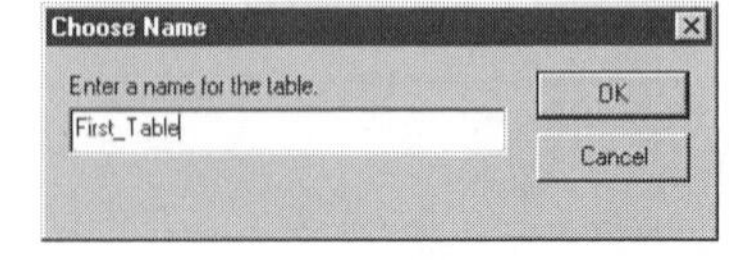

Click **OK**. You are now presented with the table design window. There are ten properties that you may need to enter for each column that you add to the table.

The asterisk (*) in the title bar of the design window indicates that there is unsaved data in the window.

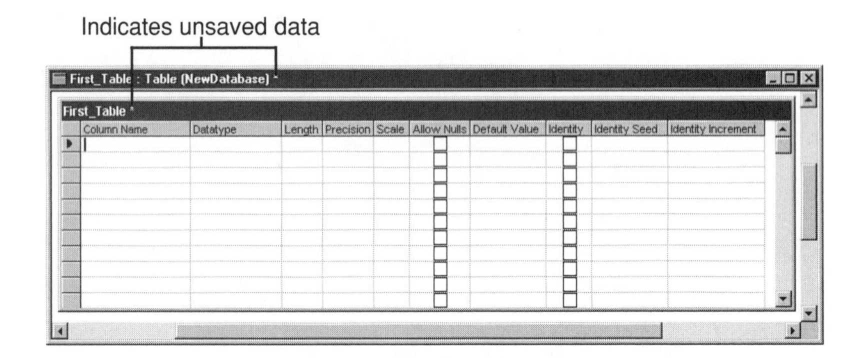

- **Column Name** A column name may be 30 characters long. It may or may not be case-sensitive depending on the settings on the SQL server.
- **Datatype** There are 20 data types in the SQL Server: char, int, currency, and datetime are a few examples.
- **Length** If a column is set to a data type of char and the length is set to 20, there may be up to 20 characters entered in the field.
- **Precision** In the decimal or currency data types, this is the number of decimal digits on both sides of the decimal point.
- **Scale** In decimal or currency data types, this is the number of digits to the right of the decimal point. Scale is never larger than the precision. Integer and floating point numbers are approximate numbers. Decimal and currency numbers are exact numbers, so the precision and scale are important properties.
- **Allow Nulls** This column is checked if the value of the column is allowed to be null.

- **Default Value** You can enter a value that is entered into a new record if no value is supplied.
- **Identity** An identity is a number that is assigned by the database engine to each new record. There will never be a duplicate number. These are often used as the key or part of the key.
- **Identity Seed** This is the starting number for an identity column. The default is 1.
- **Identity Increment** This is the amount that is added for each new identity number. The default is 1.

We created two columns for the new table. The first, named Key, is the data type of integer and is an identity column with a seed of 1 and an increment of 1. The second column is named data and has the data type of char with a length of 10.

With the table created, when we double-click the table name in the Workspace window, the table opens in an edit window. This will allow the entry of data and displays the table contents.

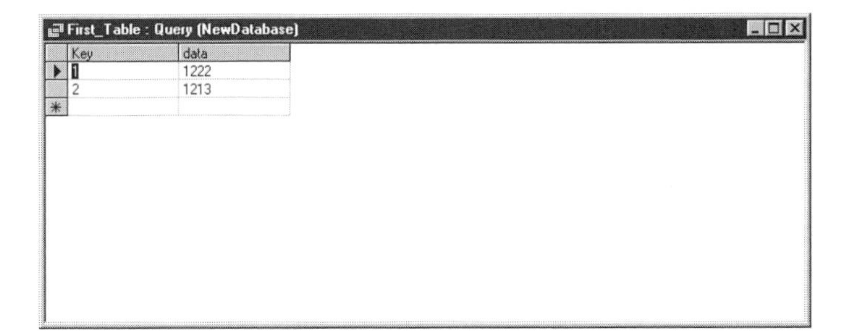

No data is entered in an identity column because the database engine will insert the value.

Creating a Database Project for an Existing Database

You can create a project for an existing database using any ODBC compliant database for which you have an ODBC driver. This can include desktop databases such as Microsoft Access, dBase, Paradox, and FoxPro. You can also work with such full strength databases as Oracle, DB2, and Microsoft SQL Server. This allows you to take advantage of existing databases without having to convert them to SQL Server or move them from their current system and platform. All you need is to be able to connect with ODBC.

Connecting to the Database

You start by choosing **File**, **New** and clicking the **Projects** tab. Select **Database Project** by highlighting it and enter a project name in the **Project name** text box. When the New dialog box looks like the next figure, click **OK**.

If the database is being added to an existing project, you start by choosing Project, Add to Project, Data Connection.

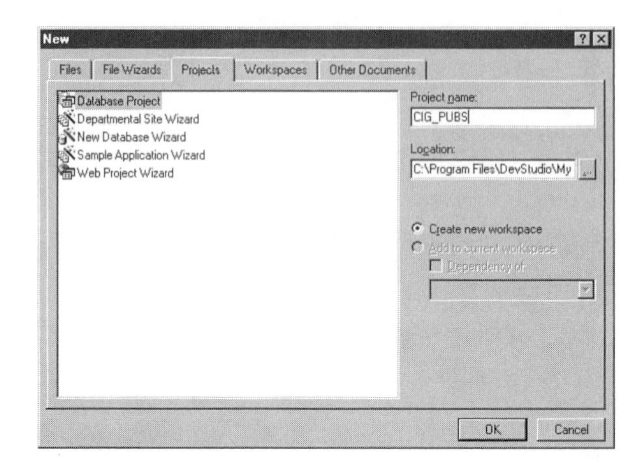

You are now asked to select a data source. Because all Web page database connections require a System DSN, you will need to click the **Machine Data Source** tab on the Select Data Source dialog box. If the data source that you need to use is shown, select it by highlighting its name. If a new data source is to be created (as it will be), click the **New** button.

Notice that the Type column shows both System and User data sources.

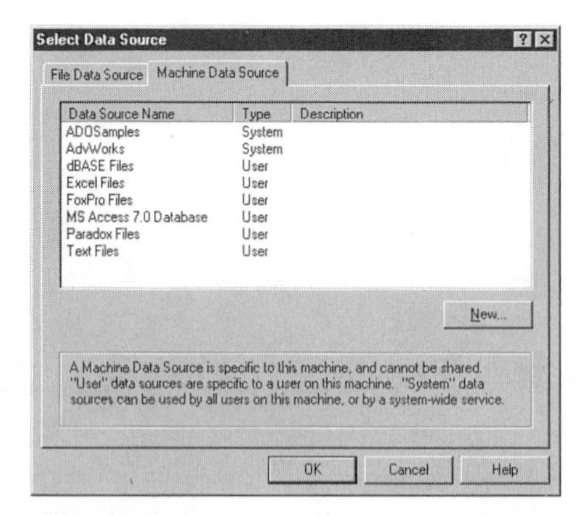

In the Create New Data Source dialog box , you must select the **System Data Source** radio button. A system data source can be seen by any application on the system and this is what you need for Web programs.

After selecting **System Data Source**, click the **Next** button. The next step in the Create New Data Source requires that the ODBC driver be selected. Highlight the name of the ODBC driver, (in this case, SQL Server) and click the **Next** button.

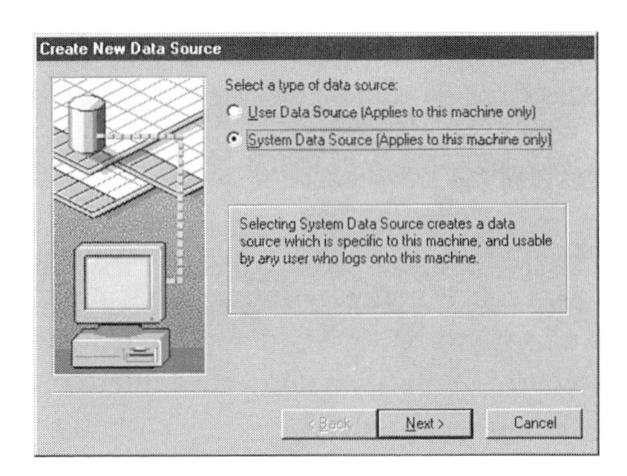

If you click the Cancel button, you will return to the previous dialog box.

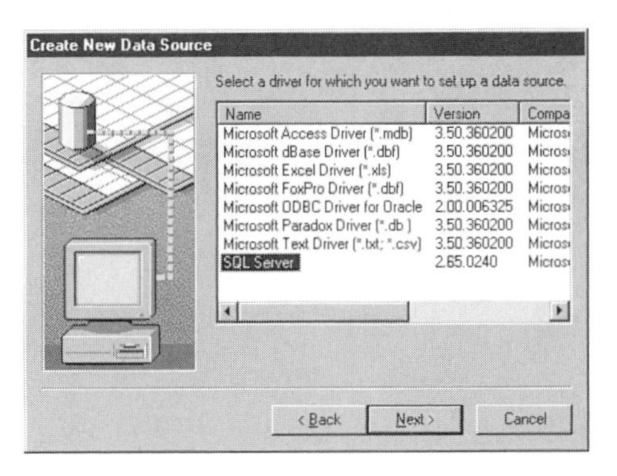

Notice the Version column. There may be times when the ODBC driver version number can be quite important in diagnosing a problem.

You are now presented with the ODBC SQL Server Setup dialog box. The first step is to click the **Options** button to open the lower panel of the dialog box. Now the Data Source Name is entered, the Server is selected, and the Database Name is entered as shown in the figure below. Click **OK** and the DSN is created.

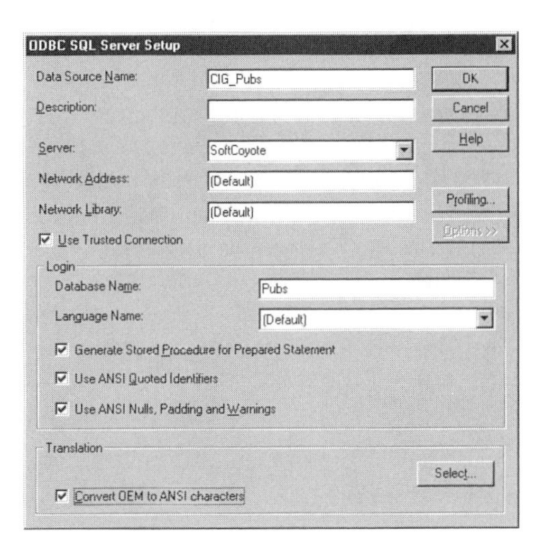

The check boxes that you should check will vary with the SQL Server setup. The database administrator is a good source to check with if you encounter problems.

You now have a functional connection to the database. Close the dialog boxes by clicking **OK** and **Close** as required.

Opening the Database

Click the **DataView** tab of the workspace window and click the open tree symbols (+) to display the database objects. When you double-click a table name, the table contents will be displayed as shown in the next figure. If you are using SQL Server 6.5, you can open the table design dialog box by right-clicking the table name and selecting **Design** from the menu.

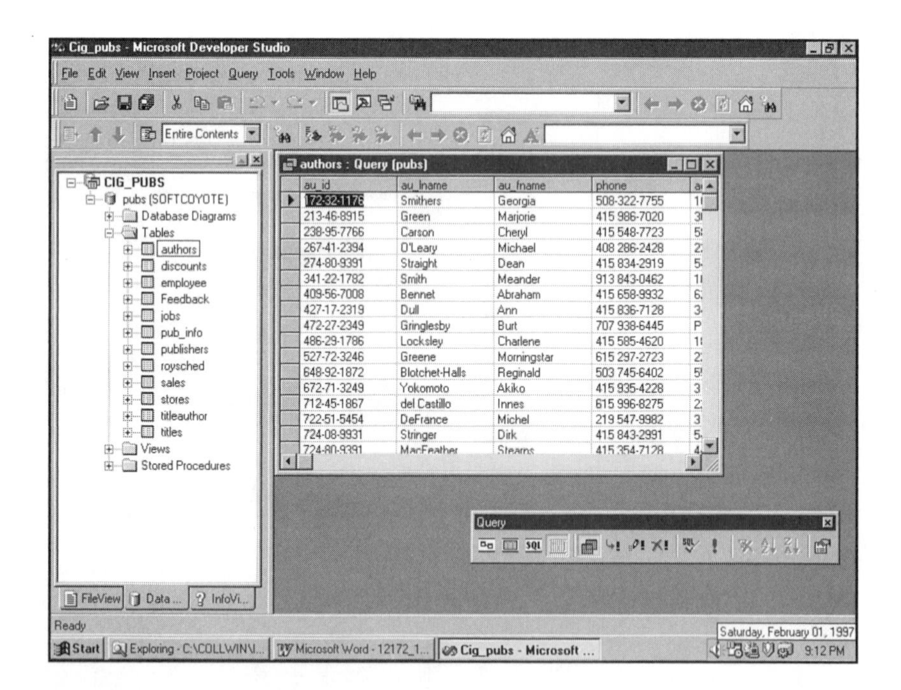

The Query toolbar also opens with the table so that you can create queries.

The Least You Need to Know

Visual InterDev provides the tools to create SQL Server 6.5 databases. It also provides the tools for connecting to most existing databases. The process is easy and quick, which allows you to concentrate on the creation of an exciting and useful Web Site rather than spending your time learning database mechanics.

- ➤ You can only create databases for SQL Server 6.5. This includes creating tables and modifying the structure of the database.
- ➤ You can use any database in your Web project for which you have an ODBC driver and a connection. Microsoft is not the only manufacturer of ODBC drivers. (They can be quite pricey but it may be worth it in terms of time and cost savings in the long run.)

Chapter 15

Database Connectivity—Long Distance, Please

In This Chapter

- Explore the use and functions of ODBC
- Learn to write VBScript to access a database using ADO

Connecting a database to a Web application provides the capability of giving users access to the latest information. You can also create full client/server applications that operate over the Internet or your local intranet. This new capability accounts for much of the shift toward Internet technology to implement client/server database applications. The database facilities in Visual InterDev make it a premier development environment for Web projects.

There are two key pieces of software that make this client/server database connection not only possible but relatively easy. The older of the two is ODBC (Open Database Connectivity). The newer is ADO (ActiveX Data Object). In this chapter we will examine these two elements.

Check This Out...

Check out the CD-ROM!
This chapter has corresponding sample files on the included CD-ROM. Simply click on "Examples," then the corresponding chapter number for which you are interested (be sure to read the Readme text file, also located in the "Examples" Section, for important information on installing the sample files).

ODBC—The Universal SQL Translator

In Chapter 12, "Relational Databases and SQL: Information, Please!," you explored the use of SQL to retrieve, update, insert, and delete data from a relational database. SQL is relatively standard from one relational database engine to another. It is one of those standards with which compliance is going to ensure that it is always the same, almost. This is an example of theory and reality diverging.

What is *not* standard is the method that the User Interface (UI) uses to communicate with the database engine. It can vary from command-line access (such as ISQL with Microsoft SQL Server through API interfaces) to an integrated UI (such as Microsoft Access). Lack of a standard interface and UI makes the choice of an RDBMS relatively irrevocable. You tend to be tied to a specific software vendor.

What is needed is a set of middleware that will stand between the user interface and the relational database engine and put the SQL statements generated by the UI into a form that can be used by the database engine.

This is exactly the task performed by ODBC. As shown in the figure below, ODBC fills this medium role of accepting SQL statements from the UI software and translating them into a form that can be used by the database engine. When the results set is created by the database engine, it is handed back to ODBC, which returns it to the UI software in a form that is readable by the UI.

ODBC acts as a bridge, allowing the database engine to be changed without changing the UI software.

SQL Statement →

← Results Set

Techno Talk

ODBC—the Database Liberator

Microsoft originated ODBC to make the UI or front-end of client/server applications independent of the back-end database. It has resulted in great flexibility in the choices of databases. It is not uncommon for an application to be designed using a desktop database such as Microsoft Access and implemented with a more powerful database such as Microsoft SQL Server. It is also excellent for using legacy databases effectively without transferring the data.

ODBC Drivers

ODBC performs this magic by using Dynamic Link Libraries (DLL) called ODBC drivers that are specific to the RDBMS being used on the back end. When the RDBMS is changed, all that must be done to provide access by the UI front-end is to change the ODBC driver that is used. The other flexibility that this provides is the capability to interface an application with multiple databases all running on different RDBMS software, and even on different computers on the network.

The UI software does not refer directly to the database. Instead it refers to a Data Source Name or DSN that is setup in ODBC. When the UI refers to the DSN, it doesn't care and is actually not aware of which RDBMS is housing the data. All the UI expects is that ODBC will return a results set containing rows of data that can be displayed and processed.

ODNC Data Source Name (DSN)

Creating a DSN is a very simple and straightforward process. You'll need to create a directory named C:\CIG_Chap16DB. To work with this example, copy the file CIG_Chap16DB.MDB (from the CD) into the directory. Then, to create the ODBC DSN (totally outside Visual InterDev), follow the steps below:

1. Choose **Start**, **Settings**, **Control Panel** and click the **32bit ODBC** icon.
2. Click the **System DSN** tab as shown in the following figure.

ODBC DSN Setup The method of creating an ODBC DSN shown here is totally outside the Visual InterDev environment. In Chapter 14, "Database Wizards, or Instant Databases," you can see how to perform this from inside Visual InterDev.

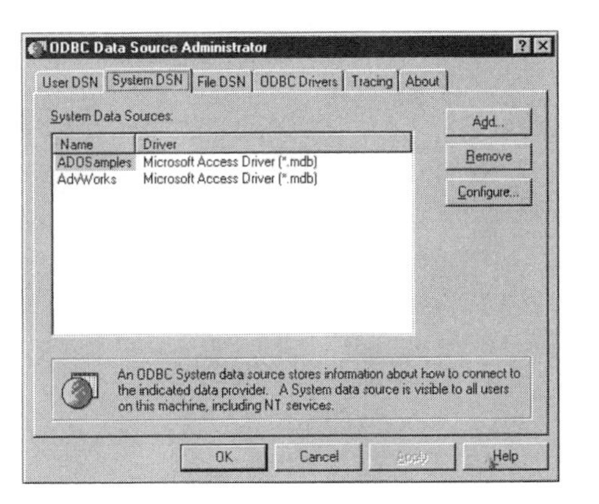

Make sure that you have selected the System DSN tab.

3. Click the **Add** button.
4. With the Microsoft Access Driver (*.mdb) highlighted (as shown in the figure below) click the **Finish** button.

All of the ODBC drivers installed on your system are displayed here.

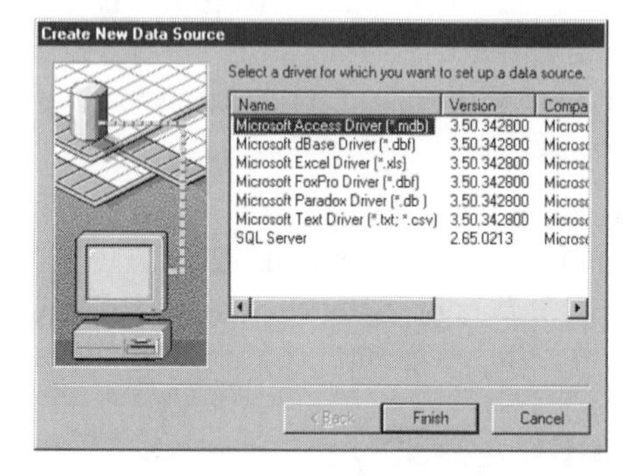

System DSN

For all of the Data Source Names that you set up to use databases with Web pages, you will need to make sure that they are always System DSN's. A System DSN is a DSN that can be seen by multiple applications on the system. If you set it up as User DSN, it will not be seen by your application. This is a common problem when connecting to databases. Check for this in case of problems while connecting to the database.

5. Enter the Data Source Name **CIG_Chap16_MyDatabase** in the appropriate text box as shown in the figure. Now click the **Select** button.

The Data Source Name can be any name.

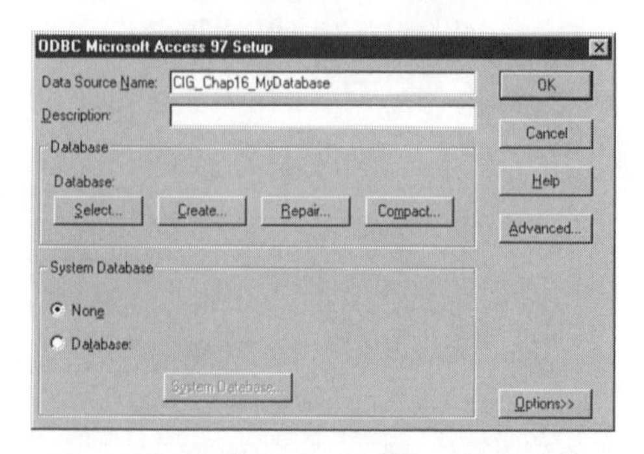

6. When you click **Select**, it opens a Select Database dialog box as shown in the figure below. Locate the directory CIG_Chap16DB, and select the **.mdb file CIG_Chap16DB.MDB** and click **OK**.

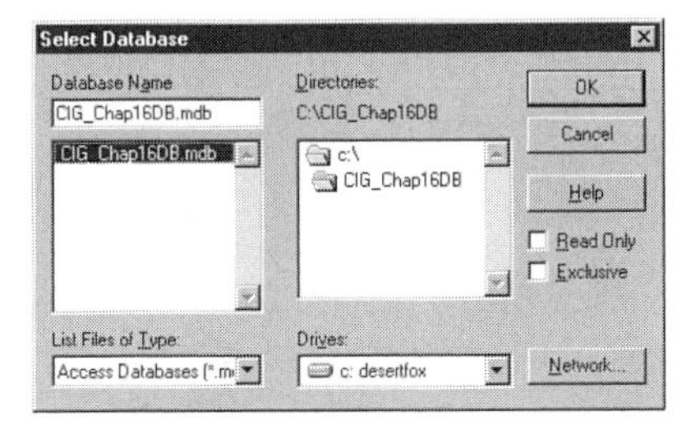

By default the Read Only and Exclusive check boxes are not checked. It is important that they stay cleared.

7. Check the **ODBC Microsoft Access 97 Setup** dialog box as shown in the following figure to be sure that the correct database is listed above the **Select** button.

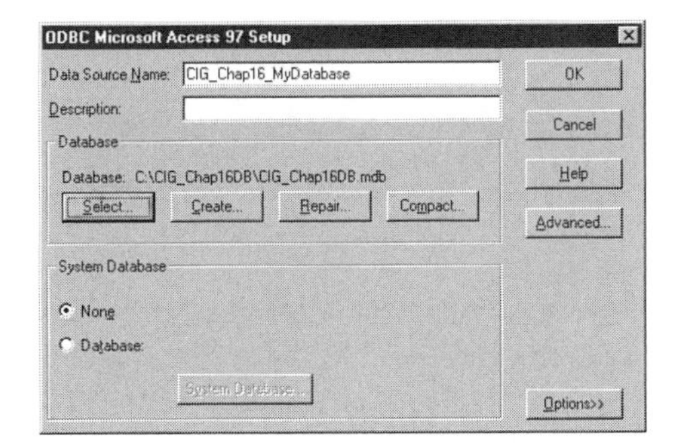

Access databases can be created, repaired, and Compacted from this dialog box.

8. Click **OK** on all of the dialog boxes as appropriate and close the **ODBC** and the **Control Panel**.

You now have a fully functional DSN that can be used by an ActiveX Data Object (ADO).

ActiveX Data Object

The ActiveX Data Object (ADO) is a set of objects that includes buffers, methods, and properties for a database connection. When you use these objects you are creating an instance of the ADO class in your application.

Single-Tier ODBC Drivers versus Multi-Tier ODBC Drivers

An important distinction in the type of ODBC driver is whether it is a single-tier or multi-tier driver. A single-tier driver functions as the database engine. There is no need to have the RDBMS software installed. The ODBC driver performs all of the work of the database engine. Examples of single-tier ODBC drivers are Microsoft Access, dBase, FoxPro, and Paradox.

Multi-tier ODBC drivers interface with database engines such as Microsoft SQL Server and Oracle. One of the largest differences besides performance and capacity is the use of stored procedures. Stored procedures are very powerful tools that control the integrity of a database, enforcing business rules, and validating data. Multi-tier ODBC drivers and the databases they support typically send less data over the communications link. This is a large performance plus when you are using relatively slow network connections over the Internet.

ADO—Using the ODBC Database

We will use the ADO (ActiveX Data Object) to connect to our database. ADO works in an Active Server Page using VBScript, so our first task is to create an Active Server page.

To begin the task, create a new project using the Web Project Wizard. Name the project **CIG_Chap16_01**. (Refer to Chapter 6, "Creating and Editing Workspaces, Projects and Files," for assistance with project creation.) When the project has been created, you should see the project in the Workspace Window as shown in the following figure.

Now create a new Active Server page by choosing **File**, **New**, highlighting the Active Server Page selection. Enter the name **CIG_Chap16_01ASP** (the file extension .asp will be automatically added by Visual InterDev), select the **Add to Project** check box and click **OK**. (You will find this file on the CD in the directory Chap_16.) When the new Active Server page is created, you will see the page open for editing.

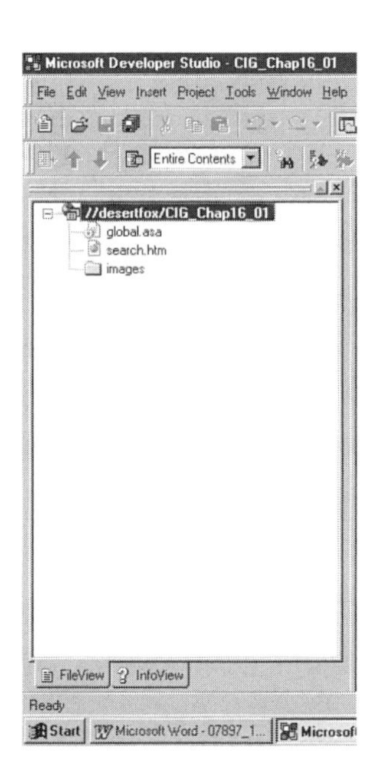

Click the plus (+) symbol and you open the project tree in File View.

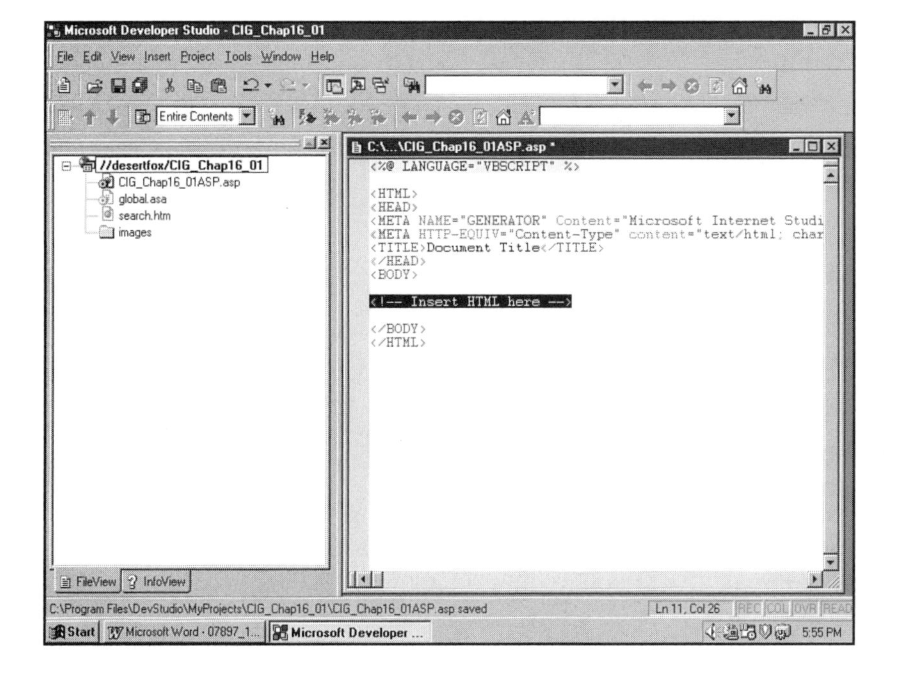

The Active Server page creation wizard has set the script language to VBScript.

Next, enter the code in the listing shown below in place of the highlighted line of code <!— Insert HTML here —>.

```
<%
     Set MyConnection = Server.CreateObject("ADODB.Connection")
     MyConnection.Open "CIG_Chap16_MyDatabase"
     Set MyRecordSet = MyConnection.Execute("SELECT * From SSN_List")
%>
<P>
<TABLE BORDER=1>
<TR>

<% For i = 0 to MyRecordSet.Fields.Count - 1 %>

<TD><B><%= MyRecordSet(i).Name %></B></TD>

<% Next %>

</TR>
<% Do While Not MyRecordSet.EOF %>
  <TR>
  <% For i = 0 to MyRecordSet.Fields.Count - 1 %>
    <TD VALIGN=TOP><%= CStr(MyRecordSet(i)) %></TD>
  <% Next %>
  </TR>
<%
  MyRecordSet.MoveNext
  Loop
  MyRecordSet.Close
  MyConnection.Close
%>
```

The edit window in Visual InterDev should now look something like the next figure.

Testing the Active Server Page

It is now time to test the Active Server page that we created. Right click the edit window and choose Preview or open your browser and set the URL to **http://servername/CIG_Chap16_01/CIG_Chap16_01ASP.ASP**. You will see the HTML page sent by the Active Server page, as shown in the figure following.

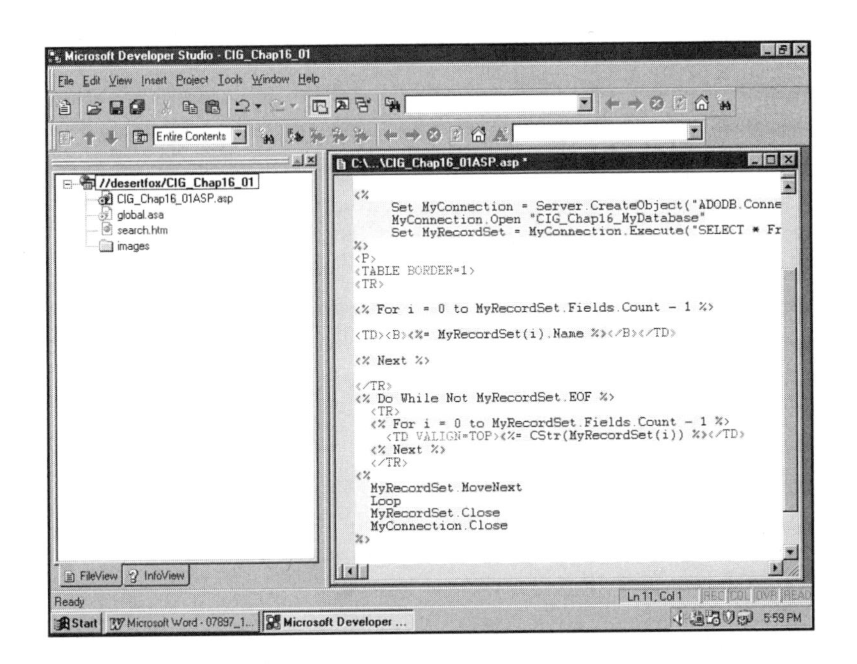

The code in the edit window is color coded according to the type. All of the VBScript is on a yellow background.

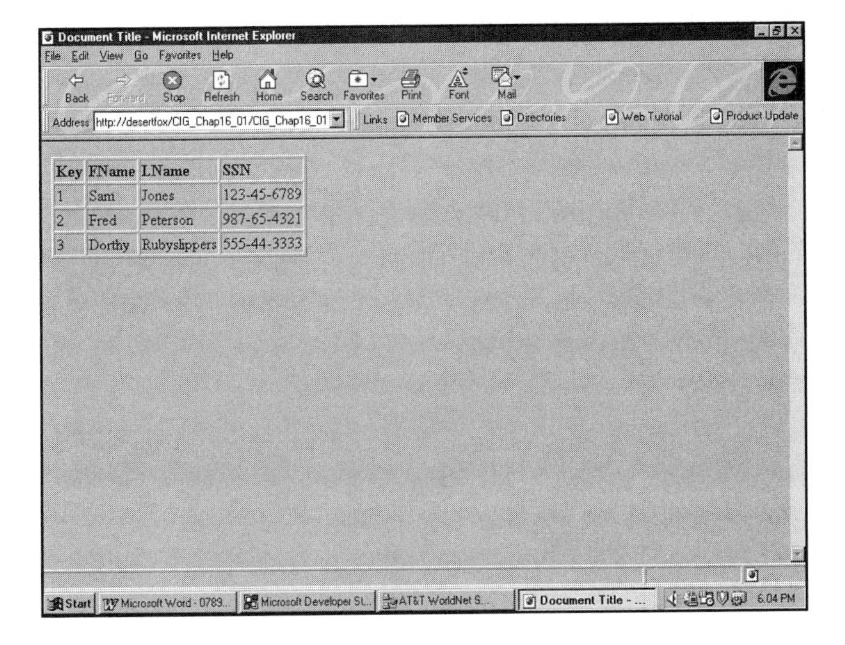

The three rows in the database are displayed in the HTML table.

Now let's look at the three lines of VBScript that establish the connection through ODBC and bring the records back to the Active Server for inclusion in the HTML page sent to the browser.

The first line creates the object MyConnection

```
Set MyConnection = Server.CreateObject("ADODB.Connection")
```

The second line uses the Open method to create the connection using the ODBN DSN that was created earlier.

```
MyConnection.Open "CIG_Chap16_MyDatabase"
```

ADO and Active Server

ADO works only with Active Server scripting. It will not work with client-side scripting because of the nature of the HTTP connection-less session. ODBC needs to communicate directly with the client software.

The third line creates the object MyRecordSet that contains the results set.

```
Set MyRecordSet = MyConnection.Execute("SELECT
* From SSN_List")
```

The remainder of the code is concerned with cycling through the record set that was retrieved from the database, and loading it into the table for transmission to the client browser for display. With these three simple lines of VBScript code, you are able to create a connection to a database through ODBC. Older methods were far more complex.

IDC and ADC

There are other methods of performing database access with Web pages.

One of these methods is the IDC (Internet Data Connector). This is being replaced in most cases by the newer ADO. It is still a good choice when you do not want to use Active Server.

The second is ADC (Advanced Data Connector). This is still in early beta testing. It promises to provide enhancements beyond the capabilities of ADO.

The Least You Need to Know

- Including database information in your Web application will greatly enhance the value of your data. One example is the ability to provide up-to-the-minute order status information to customers or sales persons over the Internet from anywhere. Complete client/server database applications are possible over the Internet or an intranet.

- All of the Internet technology database methods require the use of ODBC. Becoming familiar with ODBC will be of value.
- ODBC enables you to connect to multiple databases running on different database engines without requiring a separate interface in your Web application.
- Relational Databases are very flexible. You can add columns and views without requiring any change in the current user applications.
- ADO is the latest of a series of object-oriented database connections. New solutions will be built on a similar model. If you begin to use ADO now, then taking advantage of the latest technology will be a relatively simple matter.

Part 5
Integrating Microsoft Visual InterDev

Microsoft Visual InterDev is a very flexible and powerful tool. It provides control for most of the critical issues of Web development and maintenance.

This section will pull some seemingly scattered pieces together where you can see them as a part of the whole of your Web development toolkit.

Chapter 16

Working with Link View: Where Is That Page?

In This Chapter

- Explore the facilities of Link View for checking links
- Understand the Link View toolbar
- Understand broken links and how to fix them

Links are at the very heart of a Web site that's on the World Wide Web or an intranet. Whether the link is to another Web site, another document in the current Web site, another anchor in the current document, a sound file, or an image file, they are the fuel that runs the navigation of the Web.

A given Web site can have hundreds, even thousands of links. Keeping track of—testing—all of these links can be an impossible task. When a change is made to an HTML document, you will want to know all of the links before the change and check to see if the change has broken any links. Fortunately, Visual InterDev's Link View makes handling links easy.

Check This Out...

Check out the CD-ROM!

This chapter has corresponding sample files on the included CD-ROM. Simply click "Examples," then the corresponding chapter number in which you are interested. (Be sure to read the Readme text file, also located in the "Examples" Section, for important information on installing the sample files.)

Links Request and Load Other Files

A link is a reference to a file, document, or page. (Your Web browser requests the file to which the link points.) Some links will point to files that form part of the current page, such as image files or sound files, while some will point to other HTML documents that will be loaded in place of the current HTML document. These documents may be part of the current Web or may be in another Web site. Moreover, some links will have the function of returning you to the Web site's home page. Whatever the function of the link, if it is good, your Web page functions as you wish. If it is broken, your Web page will not deliver as you designed it.

LinkDemo Web Site

A Web site is included on the CD that will be used in the first demonstrations of Link View. The directory is LinkDemo and it contains a Default.htm file, a global.asa file, and a search.htm file. In addition, there is the Images sub-directory that contains MyMusic.mid and MyImage.gif. The directory LinkDemo is intended to be installed on the WWWroot directory. (There are also all of the FrontPage directories and files required to work with Visual InterDev.)

Using Link View

Link View can display the links for a specific object in a project, such as an HTML file or an image file. It also will explore the links of an entire Web site, whether it is on the same system or elsewhere on the Web.

If the links of a *specific* HTML file are to be explored, Link View is opened by right-clicking the file name in the Workspace window and choosing **View Links**. The Link View Toolbar appears when Link View is opened.

Choosing **View Links** opens a Link View window illustrating the links that are contained in the selected HTML file.

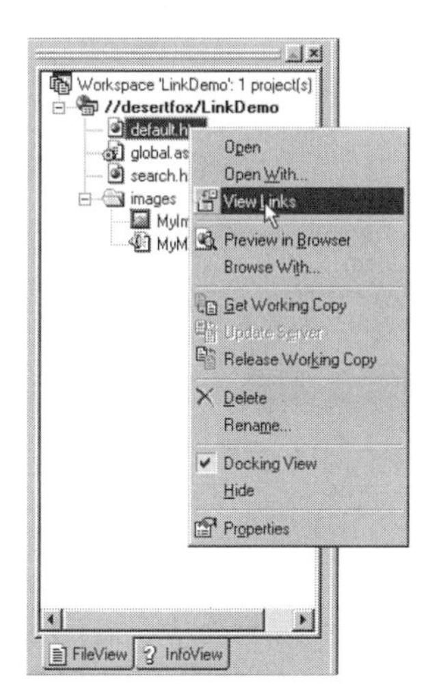

Right-clicking the file name provides access to many useful functions.

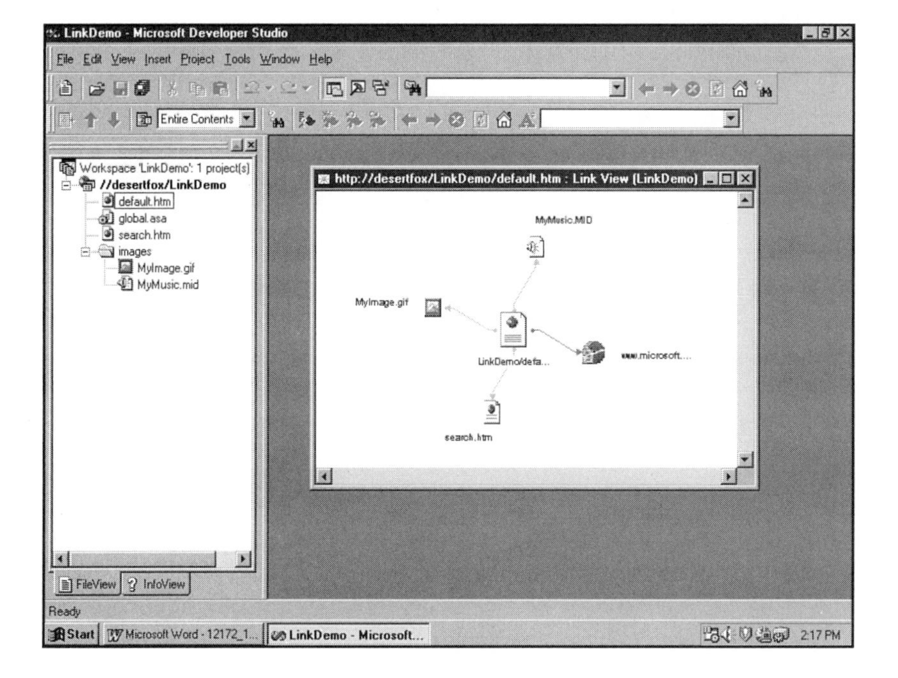

This Link View window may be printed by choosing File, Print.

When the icon of a link is shown as broken, the link was not able to be resolved. When the mouse pointer is held over the icon, the URL and any error is displayed in a ToolTip box. A link can be broken because of several factors; for example, if a file was renamed and the link still pointed to the old name. The bottom line is that Link View must follow the link. If Link View doesn't find what the link calls for, it is broken.

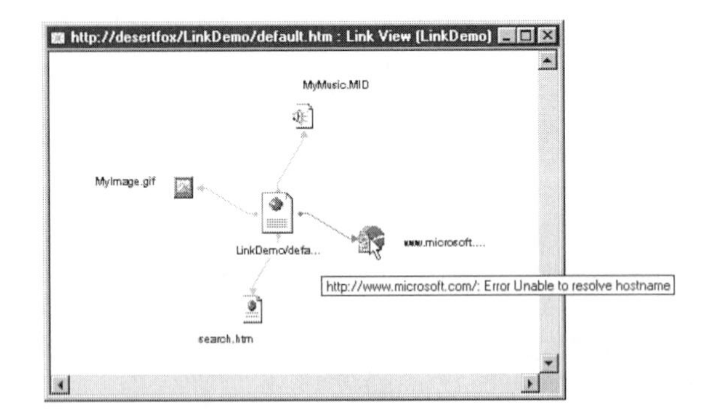

The link to www.microsoft.com was unable to be resolved because the Internet connection over a modem was not open. Notice that the icon for the www.microsoft.com is cut in half diagonally, indicating that it is broken.

If an HTML file shown in the Link View is double-clicked, the file is opened for editing in an edit window. If the file is a multimedia file, it will be opened for viewing (if an image) or listening (if a sound file).

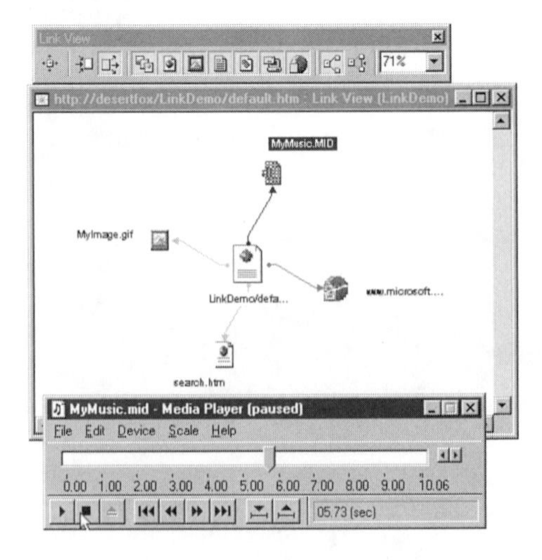

The Media Player provides the capability to listen to the sound file but has limited editing capability.

The Link View Toolbar

A look at the Link View Toolbar will illustrate the functionality of Link View. As shown in the figure below, there are 12 icons and one list box. The names of the respective icons will be shown in a ToolTip box when the mouse pointer is held over them.

The Link View Toolbar will appear when viewing a Link View Window and is hidden when the Link View Window is closed.

The names of the Toolbar tools explain the function of the tool. Opening a Link View Window and clicking the various tools illustrates their function. The tools are:

- **Expand Links** When an Object such as a Web page is selected, clicking Expand Links will show the links centered around the selected objects.
- **Show Inbound Links** This shows only links that point to the object.
- **Show Outbound Links** This shows only links from the object to another object.
- **Show All Objects** This will show links to all types of objects including images, sounds, and objects.
- **Show HTML Pages** This shows only links to HTML pages.
- **Show Multimedia Files** This shows the image and sound files.
- **Show Documents** This displays the links to files with an association to an application such as .DOC for MS Word or .XLS for MS Excel.
- **Show Executable Files** Links to files with a .DLL, .BAT, .COM, or .EXE are shown.
- **Show Other Protocols** Links to protocols other than HTTP or HTTPS are shown such as FTP.

- **Show External Files** Shows links to files and objects outside the project.

- **Show Primary Links** Shows the first level links.

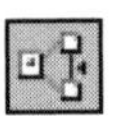

- **Show Secondary Links** Shows the second level links.

The list box is used to set the display size. This allows you to expand a List View to read details or shrink the display to see the big picture.

Looking at a Web Site

To view a Web site—whether it is on your local server or elsewhere on the Internet—choose **Tools**, **View Links** on WWW and enter the URL of the Web site as shown in the figure below.

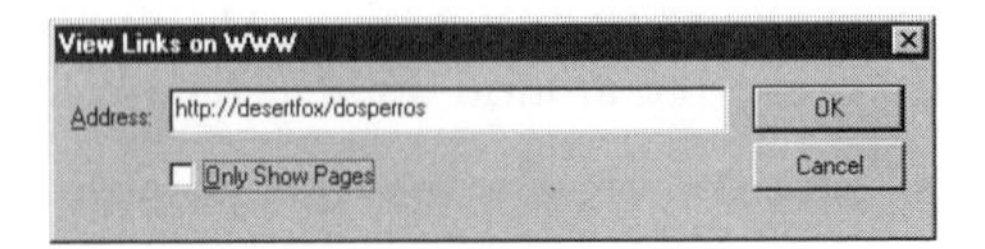

If the Only Show Pages check box is checked, only links to HTML pages will be shown.

Only Show Pages Check Box

Checking the Only Show HTML Pages check box has the same effect as clicking the same button on the Link View toolbar. If you are entering a large, complete site, you may want to leave this checked.

The Link View window shown in the figure below is the Web site of the Dos Perros sample project that was created in Chapter 6, "Creating and Editing Workspaces, Projects, and Files."

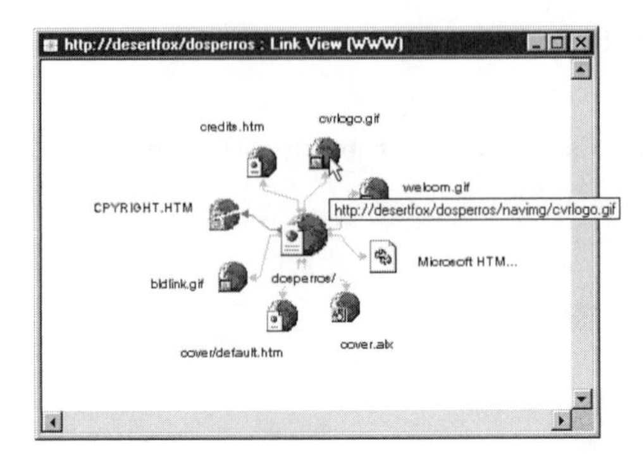

The links may be examined one at a time to ensure that it is as intended in the design.

The Least You Need to Know

Link View is a powerful design and diagnostic tool. The key to learning to use it effectively is to experiment with it. (Open a Link View of the site **http://ww.Microsoft.com** and study how the site has been constructed and the parts are linked.) By printing a LinkView, you can capture a view of the linkages on your site at a specific point in time. This can be of assistance if something gets broken later.

- ➤ Link View can be used on a single page or an entire site.
- ➤ Link View can be used to determine broken links on your site.
- ➤ Web sites on other systems can be examined with Link View. It is a great way to learn.
- ➤ There are two ways to open Link View: by right-clicking a file name, or from the menu.
- ➤ You can open the files shown in the Link View window by double-clicking them.
- ➤ Link View cannot harm a Web site. Use it often.

Chapter 17

Working with FrontPage 97: What You See

In This Chapter

- Examine the FrontPage Extensions and the functions that they perform
- See the FrontPage editor included with Visual InterDev in action
- Edit the same Web with both Visual InterDev and FrontPage

One of the excellent features that Microsoft built into Visual InterDev is the interoperability with Microsoft FrontPage. FrontPage is the Web development tool-of-choice for non-programmers and is a WYSIWYG (What You See Is What You Get) Web page editor (it has many features such as Templates and Wizards for Webs and pages). The addition of the FrontPage extensions also make the inclusion of the Search page in Visual InterDev possible since the search function is a FrontPage WebBot.

FrontPage Extensions

The FrontPage Extensions add three CGI programs to each Web that provide added functionality. These are author.dll, admin.dll, and shtml.dll. Author.dll uploads and downloads documents and updates the FrontPage ToDo List. Admin.dll sets end user, author, and administrator permissions. Shtml.dll (which stands for Smart HTML) implements the browse-time WebBot components such as Search.

FrontPage Extensions and WebBots

For use with Visual InterDev, the primary functionality added by the WebBots and FrontPage Extensions is the ability to use the FrontPage Editor, which in turn allows the use of the WebBots. The WebBots are automated functions that perform varied tasks, such as the search function that is built into some of the Webs created in this book. Other functionality provided by the WebBots include such things as the Scheduled Image Component, which allows the setting of a schedule for the display of an image in a Web page.

FrontPage Editor (Visual InterDev Version)

Visual InterDev includes a version of the FrontPage WYSIWYG Web page editor. Only files with an .HTM extension can be opened with the FrontPage Editor in Visual InterDev. To open a file with the FrontPage Editor, right-click the file name in the Workspace window and choose **Open With** from the menu. Select the Microsoft FrontPage Editor (Visual InterDev Edition) by highlighting it and click the **Open Button** as shown in the next figure. If you want to set the FrontPage Editor as the default editor, click the **Set as Default** button before you click the Open Button.

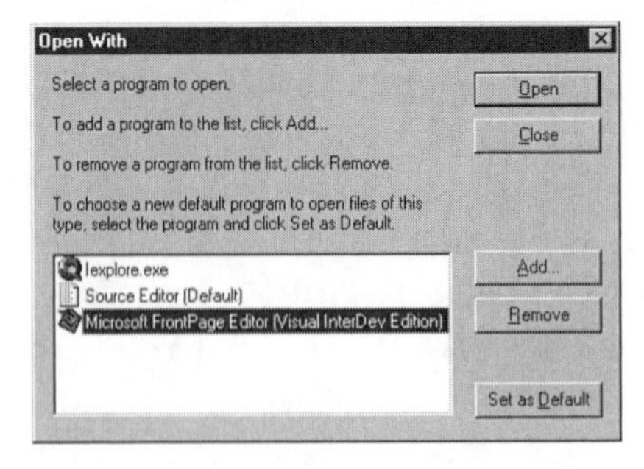

Other editors can be added from this dialog box.

The page opens in the FrontPage Editor in WYSIWYG mode as shown in the next figure.

FrontPage Editor

The FrontPage Editor allows you to create and design Web pages without working with HTML tags. There are also text and object alignment tools that look much like Microsoft Word and other Microsoft Office Applications.

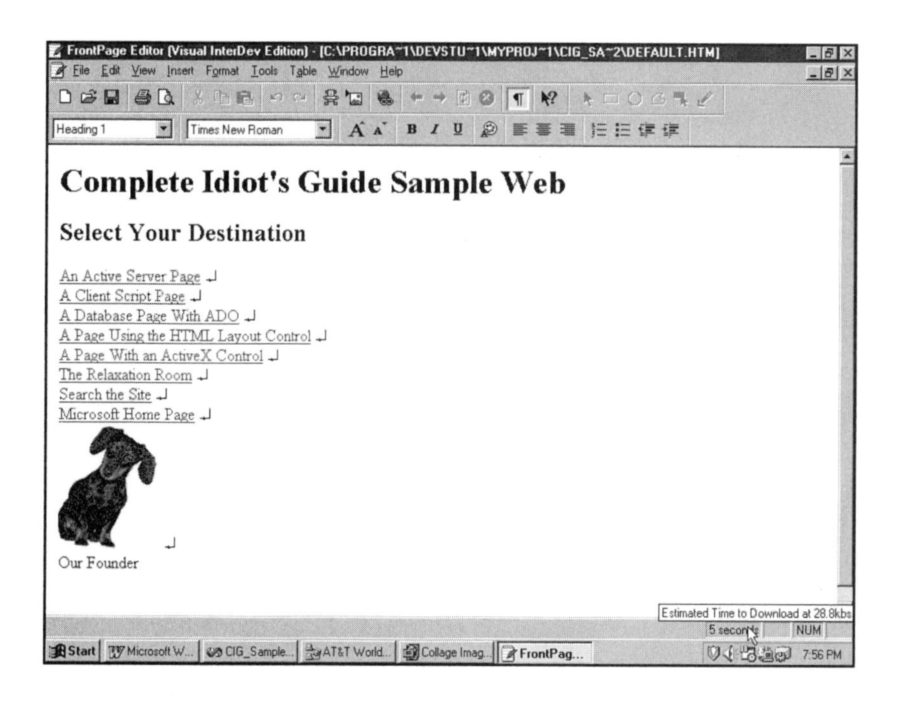

Notice in the lower-right corner of the window, the estimated download time at 28.8Kbs is given for the page.

There are many useful features in the FrontPage Editor. If you right-click the page, a menu appears (as shown in the next figure) that lets you check the Page Properties.

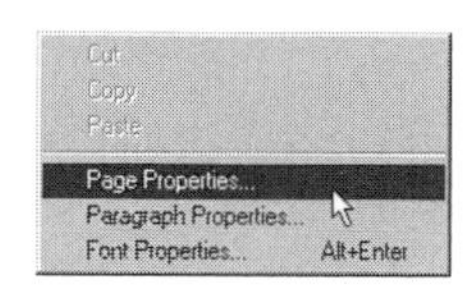

You can also check the Paragraph Properties and the Font Properties.

If you select Page Properties, the Page Properties dialog box displays the relative address of the Web page, its title, and background sound information. There are also tabs that provide additional information about the page as shown in the next figure.

Finally, if you want to see the HTML while you are editing a page in the FrontPage Editor, choose **View**, **HTML** and the HTML file will be opened in the View or Edit HTML window of the FrontPage Editor. (You have the option here of seeing the HTML color coded.)

The great advantage of having the FrontPage Editor available is that it provides an alternative view of the Web page. This can be very useful for checking details such as the background sound.

FrontPage has a very different presentation of information than does the Visual InterDev Source Editor.

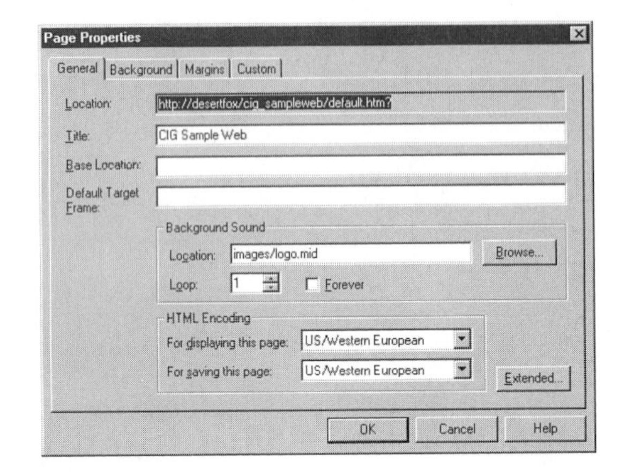

Sharing a Project Between Visual InterDev and FrontPage

The same Web can be edited using the FrontPage Explorer and Visual InterDev. This provides great flexibility for a team of developers where some members will be using Visual InterDev for the development of the scripting and database elements, while the visual layout and design members will be working with FrontPage to take advantage of its WYSIWYG features. (The full FrontPage product is being referred to here, not just the FrontPage Editor included with Visual InterDev.)

FrontPage Is Not Part of Visual InterDev

It is important to understand that the elements that are shown in this section are not part of Visual InterDev. They are part of the FrontPage 97 Web Development package and are sold separately.

When you open the FrontPage Explorer, you will be requested to select a server. You can select any server on your network for which you have sufficient access rights. After you have selected the server, click the **List Webs** button and all FrontPage Webs will be listed.

Select the Web that you wish to open and click the **OK** button. You will see the Web displayed with the links shown. In the diagram on the next page, the links read from left to right and the Default.HTM (CIG Sample Web) file is in the center. As you can see, the same file is shown on both sides of Default.HTM since there are links both ways.

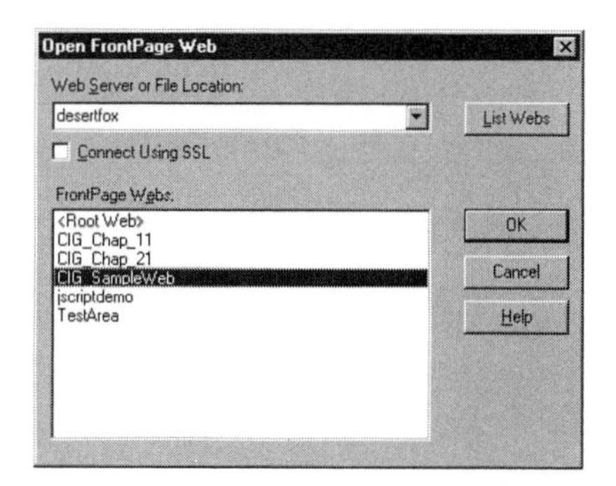

The Webs listed here were all developed using Visual InterDev.

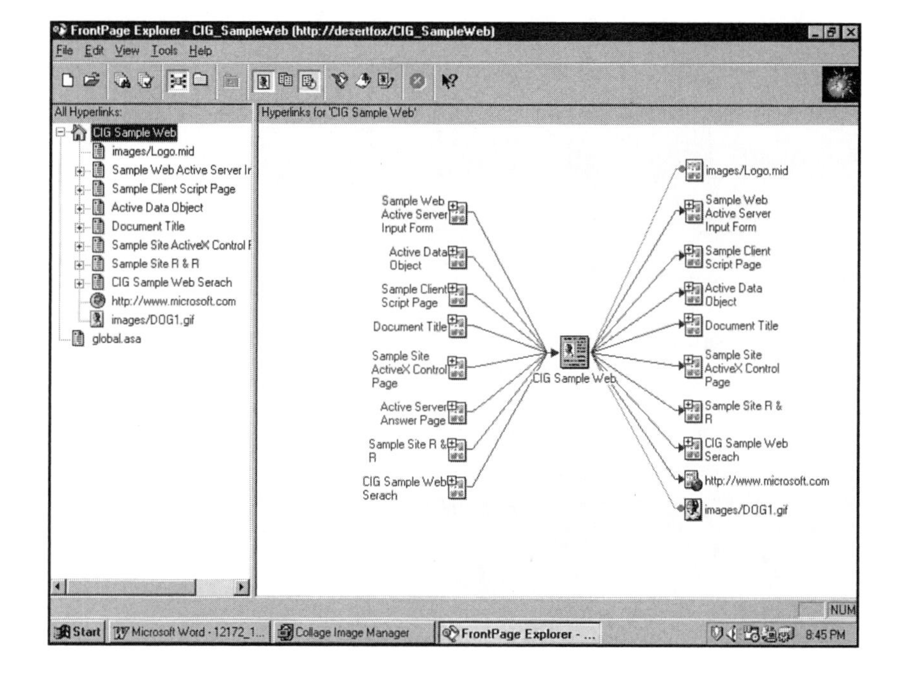

In this view, the links shown are not checked to see if they are functioning properly.

When you click the plus symbol (+) on one of the files, the links are expanded as shown in the following figure.

The CLSID is shown for any ActiveX objects.

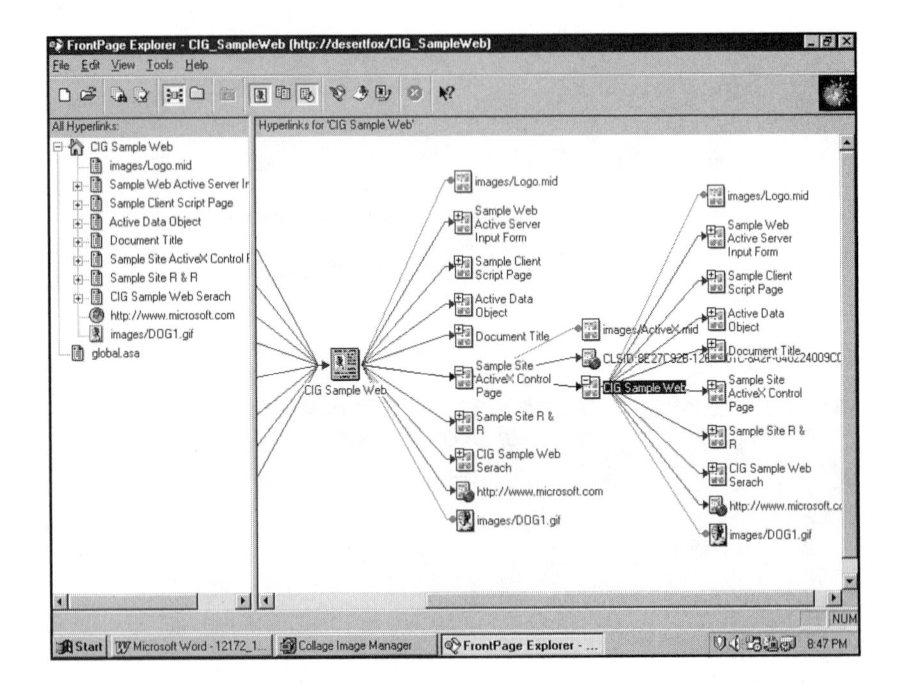

When you double-click a file icon, it is opened in WYSIWYG mode in the FrontPage Editor as shown in the next figure. This is not the same copy of the FrontPage Editor program as is opened by Visual InterDev. At the present time it is the same build of the FrontPage Editor.

This is the same edit window that is provided by the Visual InterDev FrontPage Editor.

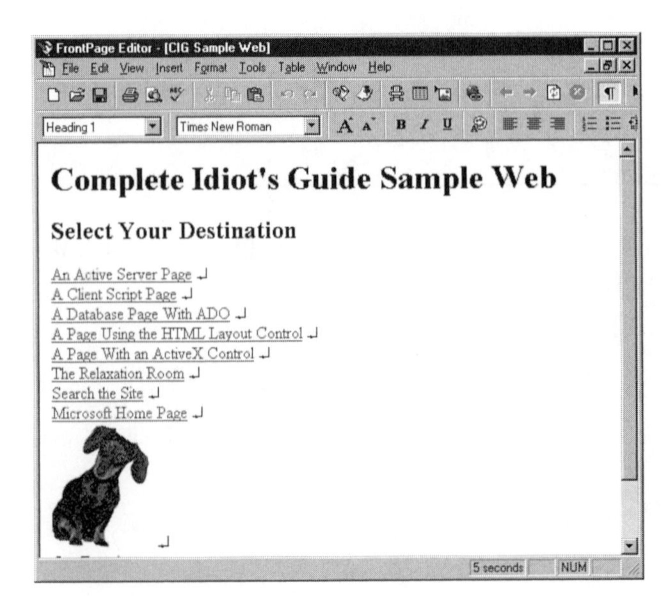

One other feature that is provided by FrontPage that is not supported by Visual InterDev is the To Do List. This is opened by choosing **Tools**, **Show To Do List** as shown in the next figure. The To Do List is a simplified mini project management tool. Tasks are added and assigned to various team members. Tasks can be marked as complete and are then either deleted or kept in a To Do History List.

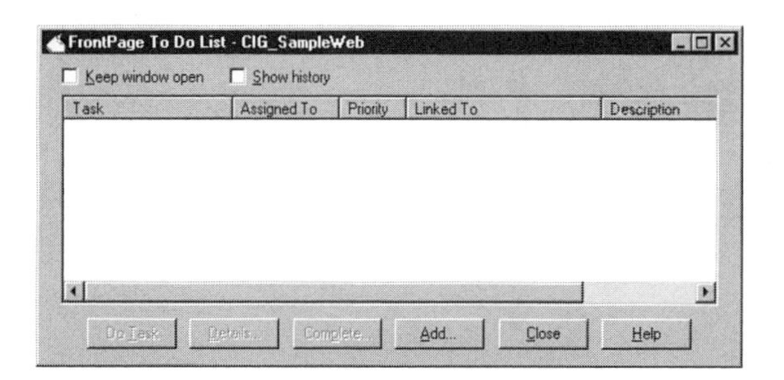

The To Do List can be seen by every member of the team who is using FrontPage.

The Least You Need to Know

The FrontPage Editor that is included with Visual InterDev provides an interesting and useful alternative to the Source editor of Visual InterDev. Projects can be jointly developed by team members using Visual InterDev and FrontPage. Just remember:

- ➤ Web projects created with either Visual InterDev or FrontPage can be edited with the tools in either development platform.
- ➤ Visual InterDev is better for working with the database and scripting aspects of the project, and FrontPage is better at the graphic aspects of the project.

Chapter 18

Source Code Security/Visual SourceSafe: You Changed What?

In This Chapter

- Install Visual SourceSafe
- Establish Projects in Visual SourceSafe
- Security in Visual SourceSafe
- Checkin and Checkout files
- Deploy a project

One of the great frustrations of team development is finding out that the project change or addition on which you have lavished time, effort, and your genius was completely lost when the change just completed by another team member was saved (or the other thrilling experience—spending hours making changes to a set of files or programs that were not the latest version). And don't forget the joy that you experience when you are unable to locate or identify the latest version of a particular file.

The conversation now lapses into, "…but I thought that you were going to…" and "…I assumed that it was my responsibility…". I have seen conversations such as these come close to physical violence. Unfortunately, not even murder and mayhem solves the problem. The problem is simple, too many cooks can spoil the soup IF THEY ARE WORKING ON THE SAME POT AT THE SAME TIME WITHOUT KNOWING IT. How do you know that someone else isn't adding ingredients to your "pot of soup" without your knowledge?

Visual SourceSafe and FrontPage

Visual SourceSafe works with FrontPage as well as Visual InterDev. The procedures for establishing a project in Visual SourceSafe are different for Visual InterDev and FrontPage. Both should be able to share the same project however. The FrontPage documentation will explain the FrontPage procedures.

Visual SourceSafe and Visual InterDev

Later in this chapter you will see that Visual InterDev and Visual SourceSafe integrate quite closely. Pay particular attention to the Shadow Folder and Deployment issues.

This is the problem that is solved by source control and security applications such as Microsoft Visual SourceSafe.

What Source Code Control Should Accomplish

The functional objectives of source code control systems are more than just keeping the latest version in a safe place. The objectives include the following:

- No code may be lost.
- No changes should be forgotten.
- No developer is out of the loop.
- Access to the code is controlled by the user.
- Previous versions can be recreated.
- Conflicting changes can be reconciled.
- Projects can be tested without impacting production systems.
- Finished projects can be deployed.

Microsoft Visual SourceSafe 5.0 meets all of these requirements and more. An entire book on Visual SourceSafe would be required to explain all of its capabilities. In the sections that follow, you will explore some of the highlights so that you will understand the essential functions.

Installing Visual SourceSafe

Visual SourceSafe has two elements that are installed, the server and the client. Both the server and the client may be installed on systems running either Windows NT 3.51 or later or Windows 95. (There should usually be only one server installation for a team of developers and a client installation on each developer's system). When you begin the setup for Visual SourceSafe, you will be presented with the choices in the following figure.

The components that may be installed are shown in the following figure. This is the setup that you will normally see only when you are installing the server.

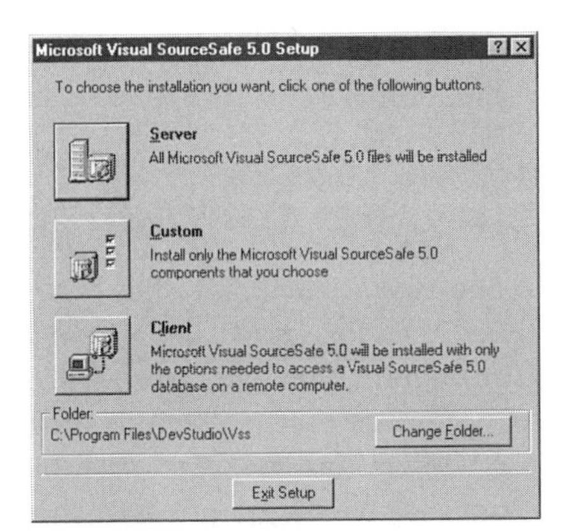

Selecting Custom will show what is being installed.

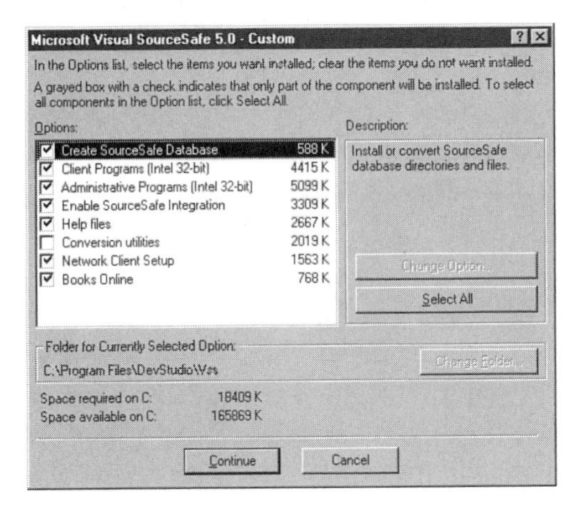

A full installation of Visual SourceSafe should only require about 20MB of disk storage.

The following are the available options when you install Visual SourceSafe 5.0:

- **Create SourceSafe Database** Creates a Visual SourceSafe database for the storage of your projects.
- **Client Programs (Intel 32-bit)** Installs a Visual SourceSafe client on the same machine as the server.
- **Administrative Programs (Intel 32-bit)** The administrative programs are used to control overall configuration of the Visual SourceSafe installation such as security and creating users.
- **Enable SourceSafe Integration** Visual SourceSafe can be integrated into various development environments such as Visual C++ and Visual Basic.
- **Help files** Of course you want the Help files.
- **Conversion utilities** These utilities provide conversion from other source control systems.

- **Network Client Setup** This allows other users to install from the Server without requiring the Installation CD.
- **Books Online** This installs a copy of the User's Guide in Help format. It is better than the book.

After installing Visual SourceSafe, the first task is to open the Visual SourceSafe Administration program. Choose **Start**, **Programs** and **Visual SourceSafe 5.0 Admin**. The first time that you open the program, you will be presented with a Message Box informing you that there is no password set on the administrator. To set the password, after the program opens, choose **Users**, **Change Password** and change the password. After you do this, each time the administrative program is opened, there will be a password challenge.

Now users must be added. (The user names can be the Windows NT domain user name. This will allow automatic user log in). Enter the User name and Password as shown in the following figure.

The same password should be used as is used for the Network logon.

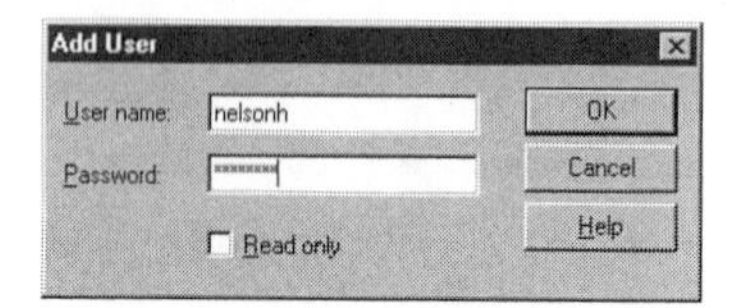

It is now possible to install Visual SourceSafe on a client system from the server. From the client system, open the directory containing Visual SourceSafe on the server. This will be \\Servername\Program Files\DevStudio\vss for a default installation. Double-click the file **Netsetup.exe**.

You are now presented with the Microsoft Visual SourceSafe 5.0 Setup dialog box. Click the **Network Client** button and the setup will be performed.

Project Security

The Visual SourceSafe security is turned on with the Visual SourceSafe Administrator program. Open the Visual SourceSafe Administrator and choose **Tools**, **Options**. Click the **Project Security** tab and check the **Enable project security** check box.

There are four access levels to the Visual SourceSafe database for domain users.

- **Read** Files may be viewed but not changed.
- **Check Out/Check In** The user may modify files.

- **Add/Rename/Delete** These users may add new files, rename files, and delete files.
- **Destroy** These users can roll back and destroy a project.

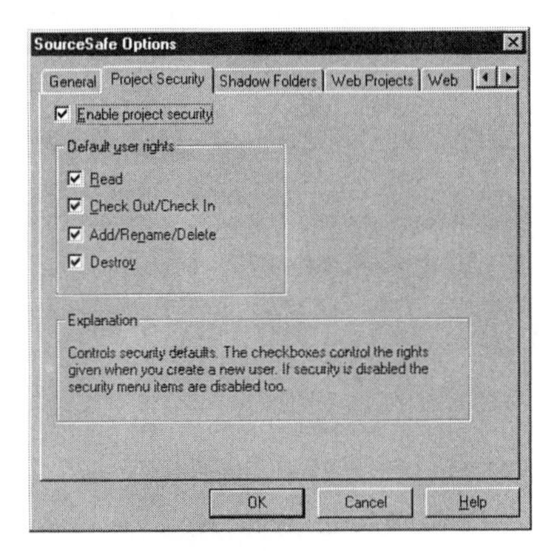

Security access rights level is established by user.

Each higher level of rights includes the other lesser rights.

Using these levels of access rights, various members of a development team can be given appropriate access to the files of the project.

Visual SourceSafe Database

The Visual SourceSafe Database is the central repository where all of the project files are stored. From the client, you will connect to the Visual SourceSafe database on the server. Open the Visual SourceSafe Explorer. Choose **File**, **Open SourceSafe Database**. Using the Browse button, locate the Visual SourceSafe database on the server and set it as the default database.

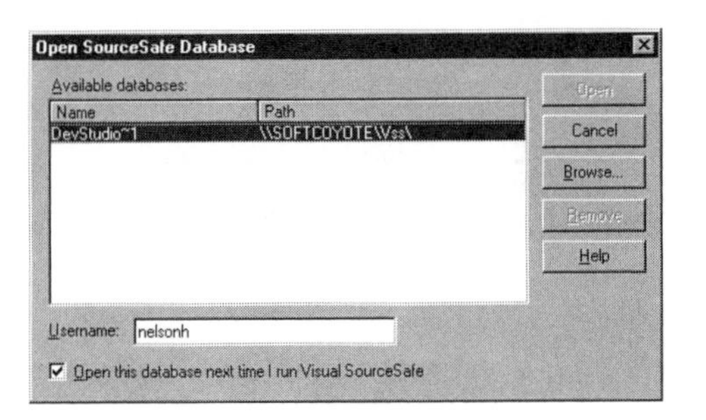

The Visual SourceSafe database always contains a file srcsafe.ini. This file contains the settings for the database.

The Working Folder

The Working Folder is the location where you will keep files that you have checked out of the Visual SourceSafe database. To set the Working Folder: With the Visual SourceSafe Explorer open, choose **File**, **Set Working Folder**. Navigate to the folder that you will use as your working folder and click **OK**.

In the example that is being used here, the files that are to be added to the Visual SourceSafe database are already in the directory that has been designated as the Working Folder. These files now need to be added to the Visual SourceSafe database. This is a very easy process. Simply open the Windows Explorer and navigate to the directory containing the files to be added to the database. Now open the Visual SourceSafe Explorer. Left-click and drag the directory containing the files to be added to the File List pane of the Visual SourceSafe Explorer.

The files are now included in the Visual SourceSafe database.

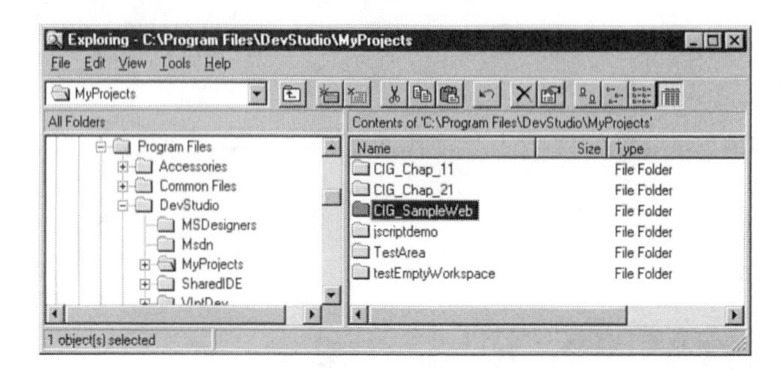

Web Projects

Visual SourceSafe is able to work with Visual InterDev. To fully utilize the Visual SourceSafe capabilities, a Visual SourceSafe project is designated as a Web project. From the Visual SourceSafe Administrator, choose **Tools**, **Options** and click the **Web Project** tab. Enter the data required as shown in the following figure.

Click **OK**. The deployment path will be discussed below in the section on deployment.

Designating a project as a Web project enables hyperlink checking and Site-Map creation.

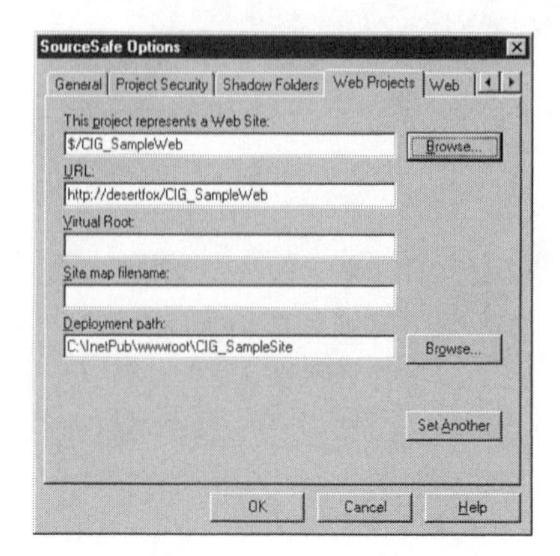

Shadow Folder

While a Web project is under construction or revision, you won't want to have the project as part of the operational Web. But in order to test the project, it must be part of the Web site. The mechanism that Visual SourceSafe provides to deal with this problem is the Shadow Folder. The project is placed in a Shadow Folder, which is a functional Web site that is used for testing. When the project is finished, it is deployed to the operational Web site. To create a Shadow Folder for a project, choose **Tools**, **Options** and click the **Shadow Folders** tab.

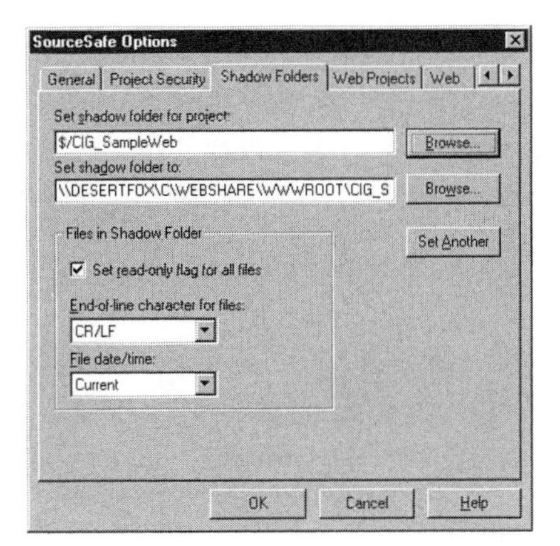

The Shadow Folder can be on another Web server.

Deploying a Project

When a project is completed and ready to be added to the production Web Site, the project is deployed. To deploy a project, in the Visual SourceSafe Explorer, choose **Web**, **Deploy**. You will see a dialog box as in the following figure. When the deployment path is correct, click **OK** and the project will be moved to the production Web server.

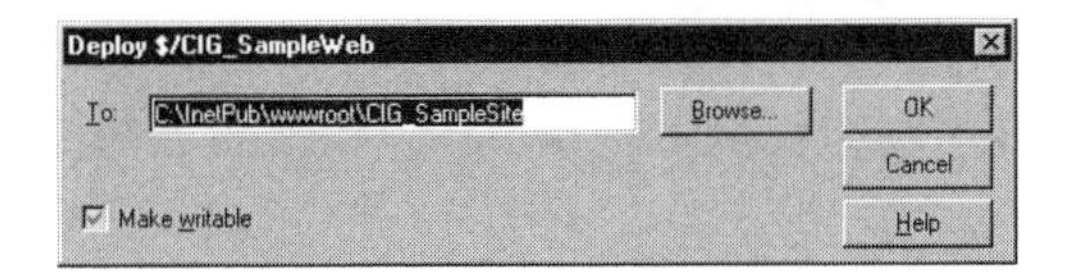

The deployment process copies the files. It does not alter any settings on the Web server.

After the project is deployed, the results are shown in the Results pane of the Visual SourceSafe Explorer.

Multiple Copies—Who's on First?

If you have been keeping count, we now have as many as 4 copies of the same file in varying conditions and versions. Each of these copies must be reconciled at some point. That is the job of Visual SourceSafe. Let's look at the location of each of these copies and how Visual SourceSafe deals with them.

- **Visual SourceSafe database** This is the secure copy. If it is changed, a history of the change is available to undo the change if necessary. This is the authoritative copy of the file.
- **Working Folder** The copy is in here while it is checked out of the database. It is used to alter and experiment with. If this copy gets deleted or completely messed up, no real harm is done since we can get a new copy from the Visual SourceSafe database.
- **Shadow Folder** This copy is in the test Web site and is here for test purposes. When the working copy is released in Visual InterDev, this is where it goes.
- **Deployment Web** This is the production Web site. The files here are only altered by deploying new versions from the Visual SourceSafe database. This insures that the production Web is always safe and reproducible in the event of disaster.

This is an excellent scheme for keeping the files and all of the past versions. If a human brain were in charge of performing this task (it would belong to a librarian), mistakes would be made.

Checking In and Checking Out

When you want to work on a file or a project, you check the necessary parts out of the Visual SourceSafe database into your working directory. You can set an option to retain a local copy of the file that is checked into the Visual SourceSafe database. To checkout a file, highlight the file name or names as shown in the next figure.

Choose **SourceSafe**, **Check Out** and the latest file copy in the Visual SourceSafe database is moved to your Working Folder. The files that are checked out are marked with the name of the user that has possession of them.

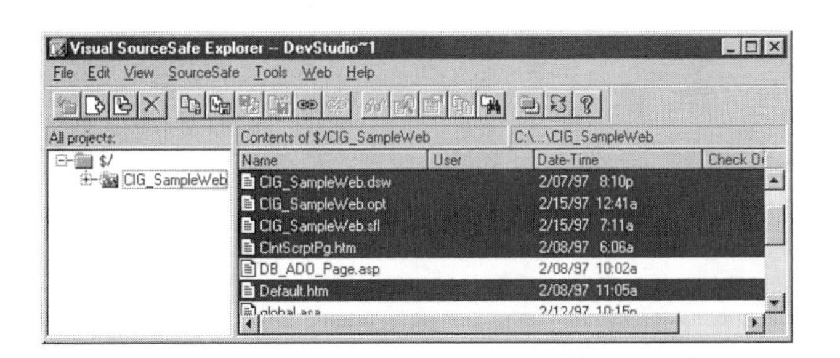

The entire project can be checked out if necessary.

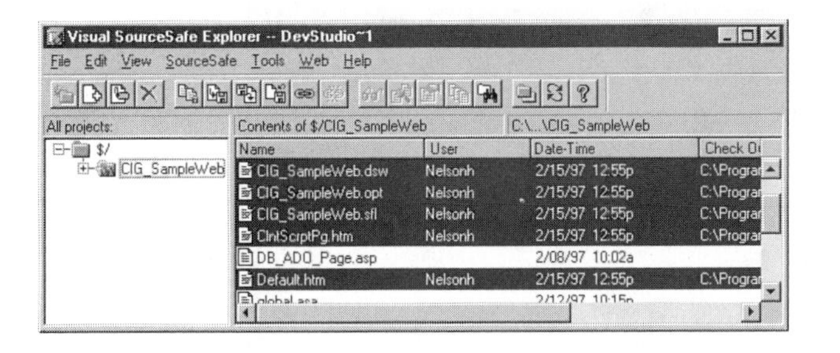

Since a file cannot usually be checked out by more than one user, showing the name allows a user in need of a file to check with the current "owner."

After you have completed your work with the files, you will need to check them back into Visual SourceSafe. Highlight the names of the files that you are going to check in and choose **SourceSafe**, **Check In**. You are asked to enter comments at the time of check in. These comments can be very useful so you should always fill them in with pertinent details.

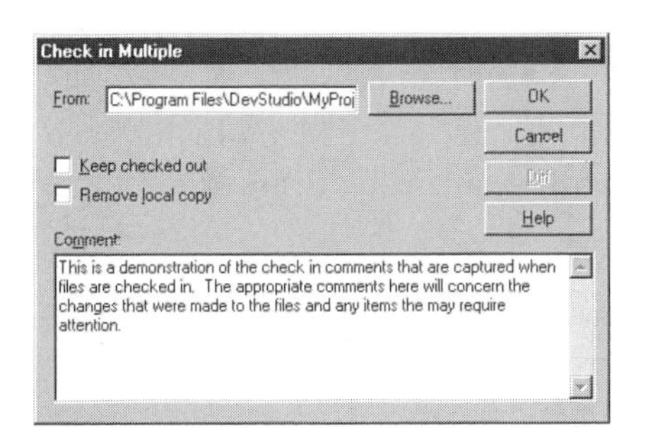

The comments can be retrieved for an individual file.

Checking Hyperlinks

To check the hyperlinks using the Visual SourceSafe tool, highlight the file or project to be checked. Choose **Web**, **Check Hyperlinks**.

Click **OK**. The results of the test are displayed in the Check Hyperlinks Results dialog box as shown in the following figure.

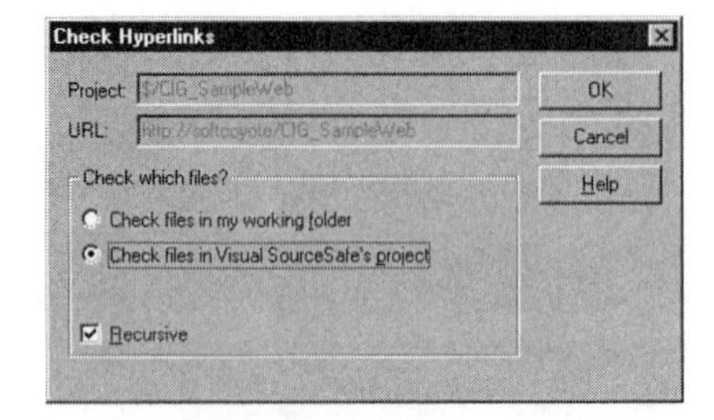

You can choose to check the files in the working folder or in the Visual SourceSafe project.

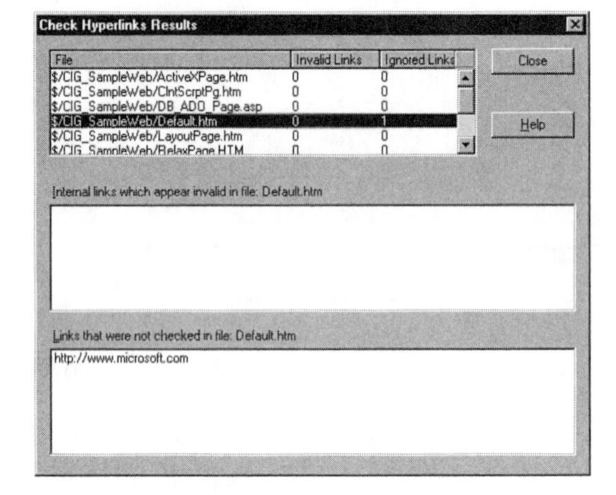

The results for individual files can be selected for display.

Creating a Site Map

Visual SourceSafe will produce a Site Map of your Web site. Highlight the project, then choose **Web**, **Create Site Map**. You are asked for a location to store the Site Map.

You can check the Site Map into Visual SourceSafe if you wish.

The Site Map is an HTML page that is opened in a browser.

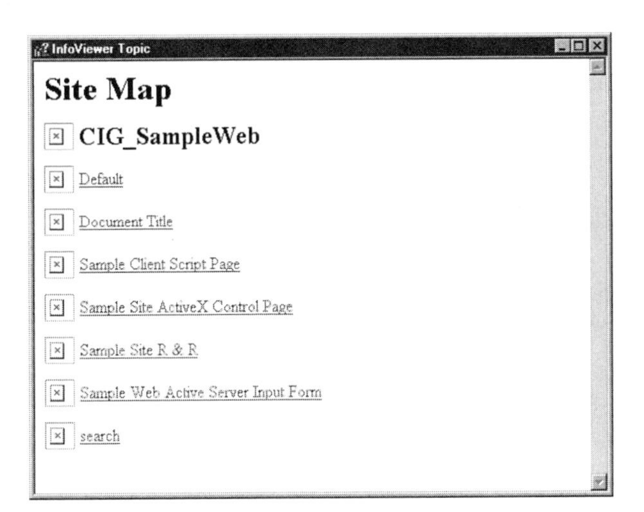

The name on each item in the Site Map is the title of the document.

Showing Project and File History

You are able to review the history of a project or file that has been controlled in Visual SourceSafe. Highlight the file or project for which you want to check the history. Choose **Tools, Show History** and the History dialog box opens.

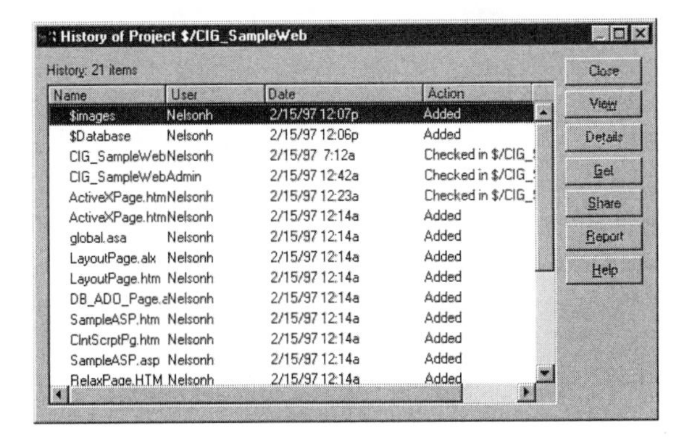

The user, date and time, and the action are shown in the summary list.

You can review the detailed comments by highlighting an entry and clicking the **Details** button. The details of the action are shown in the History Details dialog box.

You can search the list using the Previous and Next buttons.

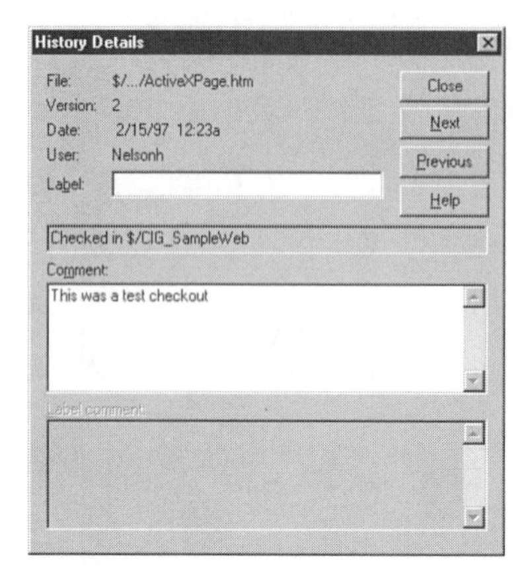

Showing Differences

Another useful tool is Show Differences. This can be the differences for an entire project or an individual file. To show the differences for a project, highlight the project name. Choose **Tools**, **Show Differences**. The differences dialog box is opened as shown in the next figure.

The differences in a project are the files and versions of files.

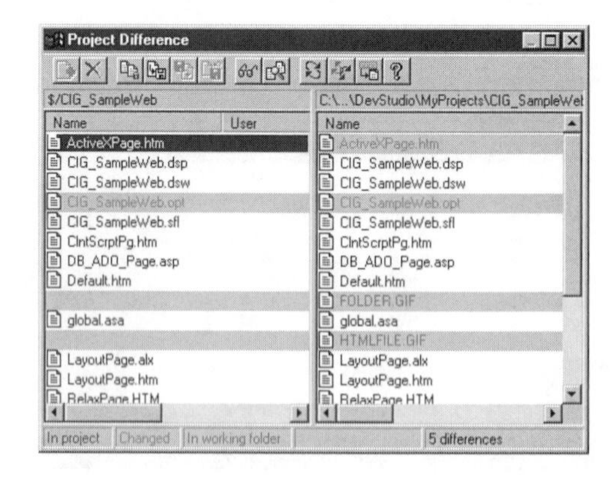

If you want to see the differences between two versions of a file, highlight the file name. Choose **Tools**, **Show Differences**. The differences are shown on a line-by-line basis.

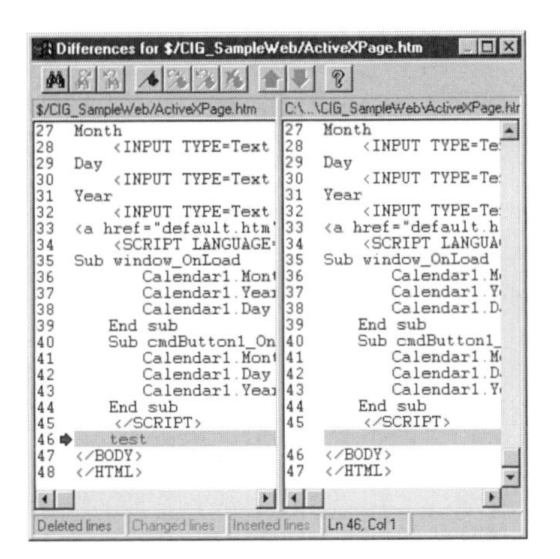

The line that is different is highlighted.

The Least You Need to Know

Visual SourceSafe is a very powerful tool. We have only scratched the surface of the features in this chapter. On large projects, there will usually be an administrator that is in charge of the policies for the use of Visual SourceSafe. Visual SourceSafe also works well in small to medium projects. (It will keep track of details that will be important to you even if you are working alone.) Remember these guidelines:

- The copy of the file that is in the Visual SourceSafe database is never worked on directly.
- The Working Folder is where files are located when they are checked out.
- The Shadow Folder is a test Web for the project.
- When the project is finished, it is deployed to the production Web site.
- If the files in the production Web site are corrupted or destroyed, the project can be redeployed.

Chapter 19

Web Site Security: Guarding the Gate

In This Chapter

- Examine the mechanisms available to secure your Web site
- Why NTFS file system for Windows NT provides the best security
- Learn why it is safe to allow access to anonymous users

Web site security seems to be a contradiction in terms. We have created a Web site and our fervent goal is to have the entire world visit and view our creation. How can we impose security on the entire world? It is always our hope that guests we invite into our home or office will be polite, well-behaved guests. The same is true of our Web site. But how can we be sure that there are not some destructive (or just clumsy) guests that will cause harm to our site?

This is where Web security mechanisms and plans come into play. The goal is to allow Web visitors access to only those parts that we want them to see and only make changes that we want them to make. (Some of the Web is "look but don't touch.") With databases, we may want them to add but not delete; change but not add; add but not look. Whatever our goals, we should be able to enforce the rules on visitors.

Our Web site security is controlled by two mechanisms, operating system security and Web Server security. The operating systems are of course Windows 95 and Windows NT. The Web servers are MS IIS and Personal Web Server.

Anonymous Access

There are two types of visitors, anonymous visitors and visitors that we know. An anonymous visitor is the typical Web surfer. The contact may be as casual as a window shopper. The visitor stops by, looks in the window for a minute or two, doesn't see anything to hold his interest (he was looking for information on antiques and our Web site covers ants) and moves on. By contrast, the visitor that we know is a member of the Windows NT domain and has a User ID and password. When he stops by, we can know that he is there because we have granted him the right to visit.

In reality, we have a User ID and password for the anonymous visitor. When a Web visitor sends a request for service to MS IIS, if we have enabled anonymous access as seen in the figure below, the IUSR_ServerName account (in this case, "softcoyote") is used to control access. You access this dialog box by choosing **Start**, **Programs**, **Microsoft Internet Server**, **Internet Service Manager** and opening the **WWW Service properties**.

The account IUSR_ServerName is created as the default anonymous account when MS IIS is installed.

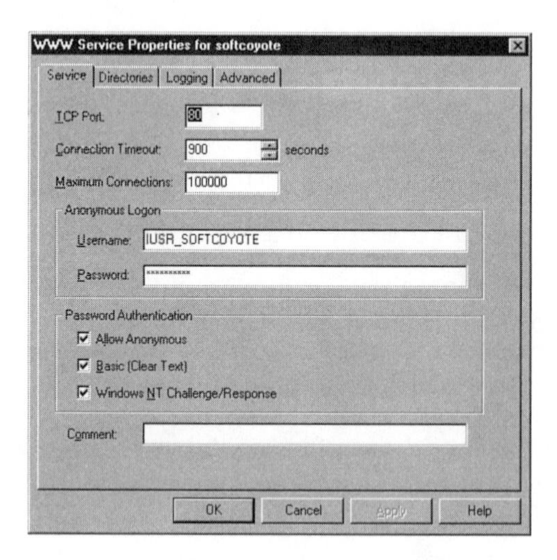

Since we have created an anonymous user and allowed this user access to the Web site, the next step is to control the access that this user has. The ultimate control of all access on a Windows NT system rests with the NT security system. The access that is granted to the anonymous user is now controlled by the Windows NT security as shown in the next figure. (If you have questions about the operation of NT security, your system administrator may be able to help.) Windows NT access control at the file and directory level requires that NTFS (NT File System) be in use on the server. By limiting the actions allowed by the anonymous user, we have moved a long way in providing security for the Web site.

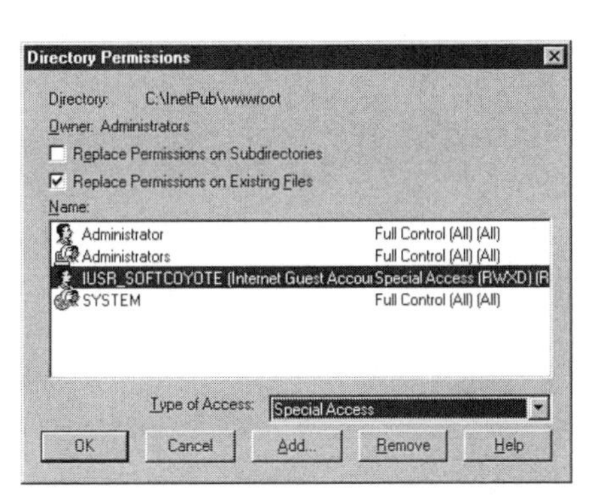

The security is based on the rights of the user. In the security here, only the anonymous user and the administrators have access to the directory WWWRoot.

Known User Access

If you can control the rights of the anonymous user, you can obviously control the access of known (User ID and Password on the NT domain) users in the same manner. The easiest method of controlling access for known users is by group.

The security provided by Windows 95 is very limited. You have the option of share level security, and for a stand-alone system, you can create a user list. In the opinion of many, Windows 95 security is an oxymoron, like a benevolent dictator.

Windows NT Security

Windows NT security is a very complex subject and far beyond the space available here. If you need assistance in this area, see your network administrator. A significant part of her job is worrying about these issues and knowing what rules to enforce and how to enforce them.

Root Directory, Virtual Directories, and Directory Browsing

The directory structure of the Web provides a large degree of control over the Web visitor. The Web visitor cannot access any directory that is not defined in the Web directories list. If a Web client browser sends a request in the form of **http://www.yourdomain.com** without a file name, it will come in at the <Home> directory, as shown in the next figure.

Multi-homing a server allows for virtual servers, each with only one home directory.

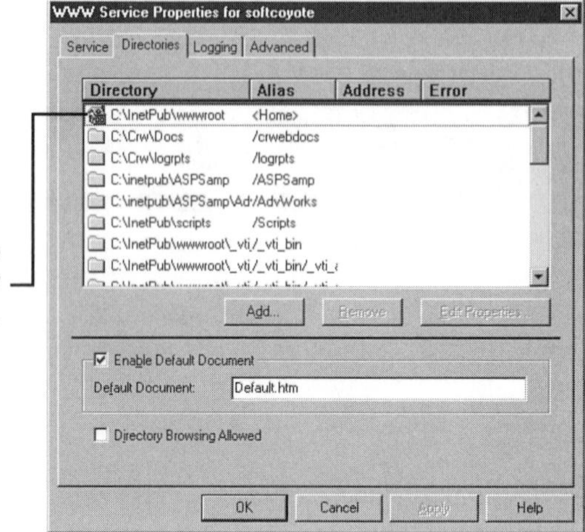

There can be one home directory per server.

This is provided that there is a default document in the home directory. If there is no default document in the home directory, the result is the HTTP /1.0 403 Access Forbidden message.

The result can be different if directory browsing is enabled. In the next figure, directory browsing is enabled. Notice that the directory structure is that of the Web server and not the physical hard drive.

No Directory Browsing Allowed

The message here is: don't enable directory browsing unless you have good reason, and have thought through the consequences.

For a user to enter a directory other than the root directory of your site, they will need to know the alias of the directory and the name of a file in that directory (if there is no default directory and directory browsing is not allowed).

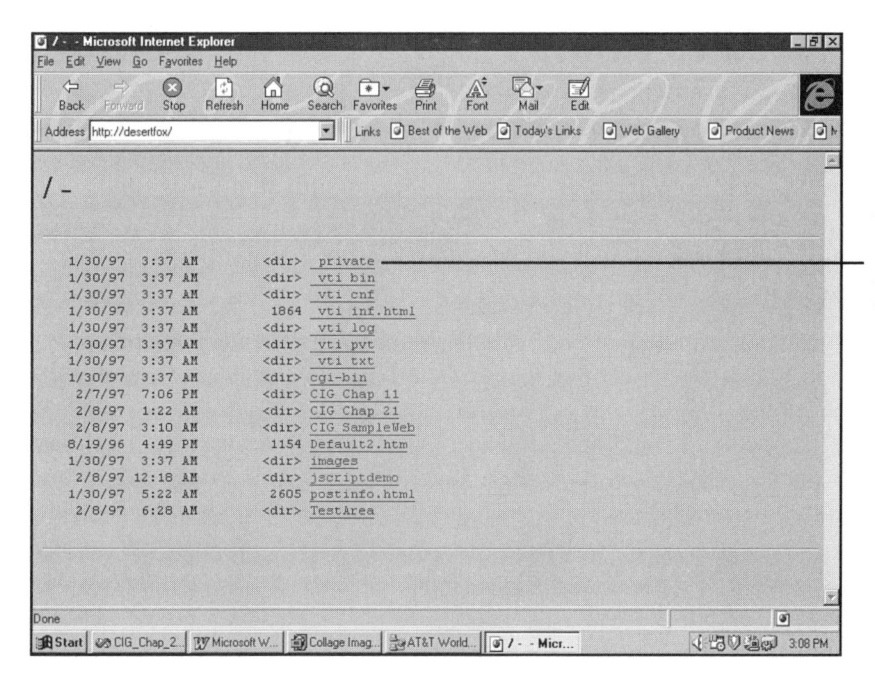

When directory browsing is enabled, your visitor may see more than you intended.

Read and Execute Permission

One final element can provide some security. If you have carefully crafted a Web application, you are not anxious to have anyone with a Web browser get their hands on your work. If you place your executables in directories with only Execute access enabled, they will be able to run but cannot be looked at. The figure below shows that you have the option of allowing Read access, Execute access, or both on a directory.

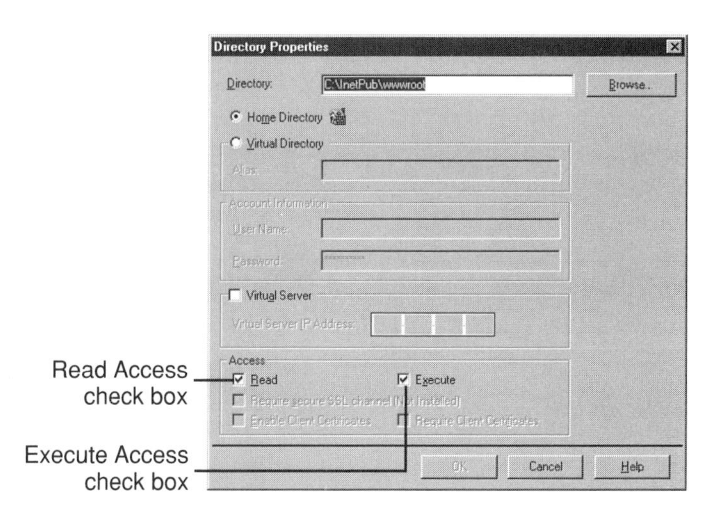

Directories are usually either Read or Execute. Rarely will the same directory be both Read and Execute access.

The Least You Need to Know

There are several mechanisms that are involved in Web site security. It is a combination of Windows NT security and MS IIS parameters that will provide protection. Just remember:

- Web site security is a serious subject. If you are not thoroughly familiar with all aspects of security, get your system administrator involved as an advisor. If you are extensively familiar with security, get someone to review your security plan to see if there are any holes in it.
- Windows 95 security is not adequate to provide significant protection. Friends don't let friends rely on Windows 95 security.

Chapter 20

Client Scripting—No, You Do It

In This Chapter

- ➤ Explore several of the HTML intrinsic controls
- ➤ Create VBScript to use the controls
- ➤ Create and examine examples

Check out the CD-ROM!

This chapter has corresponding sample files on the included CD-ROM. Simply click "Examples," then the corresponding chapter number for which you are interested. (Be sure to read the "SampleReadme" text file, however, for important information on installing the sample files.)

It is possible to perform interesting and useful processing on the client side of the client/server equation. In this case the server performs the role of sending files in response to requests by the client.

What Can Client Scripts Do?

While ActiveX Controls and Java applets present very interesting and powerful uses, you can accomplish much using the intrinsic HTML Form Controls and VBScript. Let's examine some of the intrinsic HTML Form Controls and see what we can do with them.

Button Control

The Button control is the control that you will provide the user to initiate events. The Button control generates an OnClicked event that is used by VBScript to process a block of instructions.

The Button control "Button 1" is created by HTML.

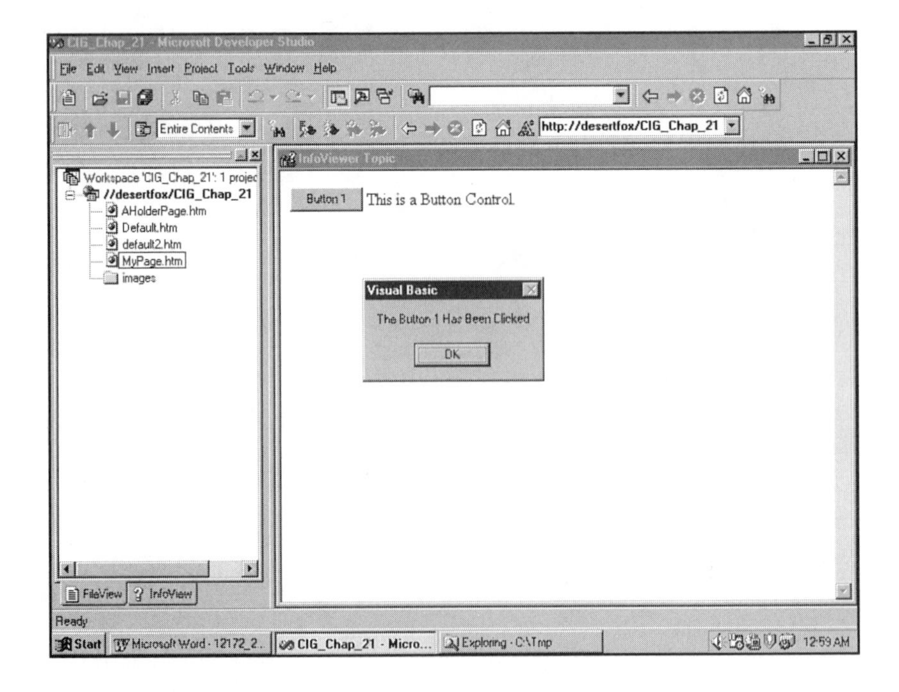

A look at the HTML for this page will show what is happening:

```
<HTML>
<HEAD>
<META NAME="GENERATOR" Content="Microsoft Developer Studio">
<META HTTP-EQUIV="Content-Type" content="text/html; charset=iso-8859-1">
<TITLE>Document Title</TITLE>
</HEAD>
<BODY BGColor =#FFFFFF>
<INPUT TYPE="BUTTON" NAME="cmdButton1" VALUE="Button 1">This is a Button
Control.<br>
<SCRIPT Language="VBScript">
   Sub cmdButton1_OnClick
     MsgBox "The Button 1 Has Been Clicked"
   End sub
</SCRIPT>
</BODY>
</HTML>
```

The following line of code does a few things:

```
<INPUT TYPE="BUTTON" NAME="cmdButton1" VALUE="Button 1">This is a Button
Control.<br>
```

- It creates the Button control as an HTML intrinsic control.
- The TYPE argument tells the type of control being created.
- The VALUE argument provides the caption that appears on the face of the button.
- The NAME argument "cmdButton1" is used to identify the control in the subsequent VBScript.

In this VBScript, a procedure Sub cmdButton1_OnClick is created to use the `OnClick` event that is generated by the Button control to create a Message Box.

```
<SCRIPT Language="VBScript">
   Sub cmdButton1_OnClick
     MsgBox "The Button 1 Has Been Clicked"
   End sub
</SCRIPT>
```

Reset Control

The Reset control is a specialized HTML Button control that returns all of the other controls on the form to their values when loaded. No VBScript is required to operate this control. It does produce an OnClick Event that can be used to do other functions. To see this control's primary use, when the page has a Text control, type a few characters into the Text control, then click the Reset control and see that the Text control is returned to the original state. The Reset control is created with an HTML line like this:

```
<INPUT TYPE="RESET" NAME="Reset1" VALUE="Clear">This is a Reset Button
Control.<br>
```

This line will produce a control as shown in the following figure.

The code in the page now has a Sub for the handling of the `OnClick` event for the Reset control. Even with the event handler it will still perform its function of resetting the other controls.

The Message Box shown is a modal window. This means that no other action can be taken in the application until it receives a response.

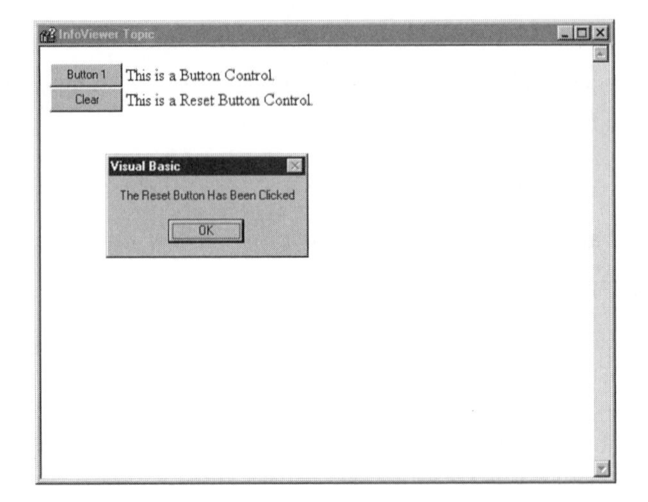

Submit Control

The Submit control is another specialized HTML Button Control that serves the function of posting data from a form to an Active Server Page. We examined this control in Chapter 8, "VBScript: Lights, Camera, Action!" It also generates an OnClick event that can also be utilized for other procedures in an event handler, as was done with the Reset control.

Text Control

The Text control is a text box that is used to enter or display data. It is created with the HTML code shown below.

```
<INPUT TYPE="TEXT" NAME="Text1" SIZE="30">This is a Text Control.<br>
```

The NAME property is used to identify the control. The contents of the Text control are contained in the `Value` Property. The code snippet below has been extracted from the VBScript of the page. It show an event handler for Button 1 that sets the value of the Text control.

```
Sub cmdButton1_OnClick
   Text1.Value="Button 1 Has Been Clicked"
End sub
```

The results of this event handler are shown in the following figure.

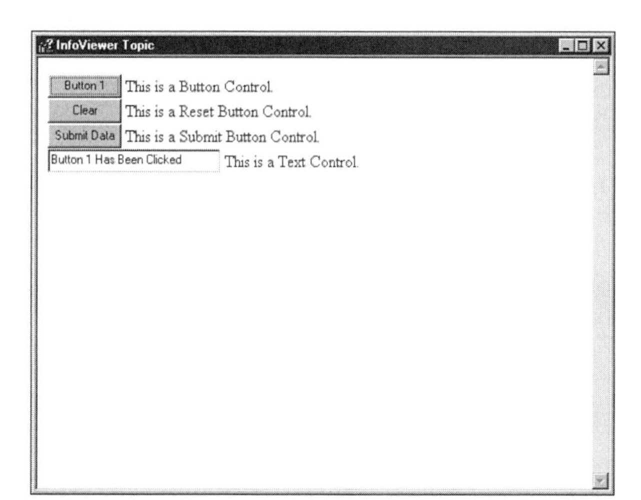

The Text control may be used for either input or output.

Password Control

The Password control is a specialized Text control that has the characteristic of displaying data that is entered as asterisks (*).

Radio Button Control

The radio button control is also called an option button. Radio buttons usually come in groups of two or more. They are used to provide the user with a choice of one out of N number of options. The choices are always only one of the number of possible choices. For example, Male or Female; Large, Medium, or Small; Hot or Cold. A person is never both Male and Female. A soda is never both Large and Medium.

Some points on radio buttons:

- Radio buttons are thought of as being in groups.
- It is possible to have more than one group of radio buttons on a page.
- The method of determining a group of radio button controls is by the Name property of the control.
- All the radio buttons in a group will have the same name:

The Code snippet below illustrates this. There are two groups of radio buttons:

```
<INPUT TYPE="RADIO" NAME="OptRadio" VALUE="Select_1">Pair 1 Selection
1<br>
<INPUT TYPE="RADIO" NAME="OptRadio" VALUE="Select_2">Pair 1 Selection
```

```
2<br>
<hr>
<INPUT TYPE="RADIO" NAME="OptRadio2" VALUE="Select_1">Pair 2 Selection
1<br>
<INPUT TYPE="RADIO" NAME="OptRadio2" VALUE="Select_2">Pair 2 Selection
2<br>
```

The radio button control has an OnClick event that can be used to activate an event handler procedure. Only one of the radio buttons in a group can be checked as shown in the figure below.

The Radio Button Control is useful when you need to force a user to make a choice.

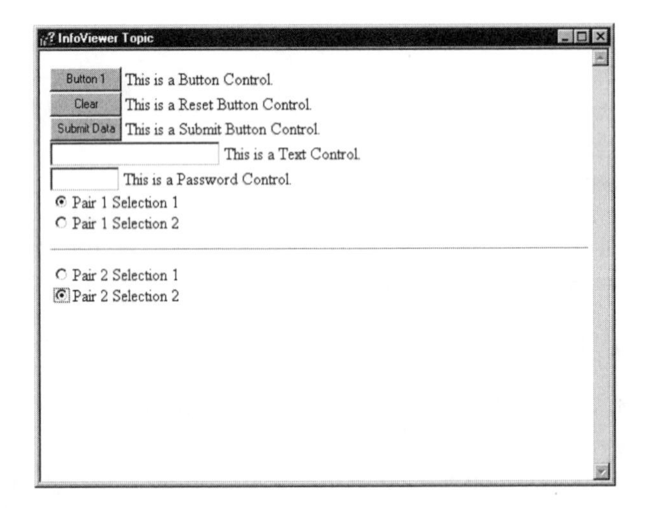

Check Box Control

The Check Box control is useful when multiple choices may be made from a list. Check Box controls are created in the code snippet below.

```
<INPUT TYPE="CHECKBOX" NAME="CHECK1" VALUE="Checked 1">Checked 1
<INPUT TYPE="CHECKBOX" NAME="CHECK2" VALUE="Checked 2">Checked 2
```

The Checked property is used to determine if the Check Box control is checked. An example is in the VBScript code snippet below.

```
Sub cmdButton1_OnClick
   If Check1.Checked then
   Text1.Value="Button 1 Has Been Clicked While Check 1 is Checked."
   End if
End sub
```

The result of this is shown in the following figure.

Testing the checked property of Check Box controls is an excellent way to capture input from the user in a predetermined form.

The full listing of the page is shown below:

```
<HTML>
<HEAD>
<META NAME="GENERATOR" Content="Microsoft Developer Studio">
<META HTTP-EQUIV="Content-Type" content="text/html; charset=iso-8859-1">
<TITLE>Document Title</TITLE>
</HEAD>
<BODY BGColor =#FFFFFF>
<INPUT TYPE="BUTTON" NAME="cmdButton1" VALUE="Button 1">This is a Button
Control.<br>
<INPUT TYPE="RESET" NAME="Reset1" VALUE="Clear">This is a Reset Button
Control.<br>
<INPUT TYPE="SUBMIT" NAME="Submit1" VALUE="Submit Data">This is a Submit
Button Control.<br>
<INPUT TYPE="TEXT" NAME="Text1" SIZE="50">This is a Text Control.<br>
<INPUT TYPE="PASSWORD" NAME="Pass1" SIZE="10">This is a Password
Control.<br>
<INPUT TYPE="RADIO" NAME="OptRadio" VALUE="Select_1">Pair 1 Selection
1<br>
<INPUT TYPE="RADIO" NAME="OptRadio" VALUE="Select_2">Pair 1 Selection
2<br>
<hr>
<INPUT TYPE="RADIO" NAME="OptRadio2" VALUE="Select_1">Pair 2 Selection
1<br>
<INPUT TYPE="RADIO" NAME="OptRadio2" VALUE="Select_2">Pair 2 Selection
2<br>
<INPUT TYPE="CHECKBOX" NAME="CHECK1" VALUE="Checked 1">Checked 1<INPUT
TYPE="CHECKBOX" NAME="CHECK2" VALUE="Checked 2">Checked 2
<SCRIPT Language="VBScript">
   Sub cmdButton1_OnClick
     If Check1.Checked then
        Text1.Value="Button 1 Has Been Clicked While Check 1 is Checked."
     End if
   End sub
```

```
    Sub Reset1_OnClick
      MsgBox "The Reset Button Has Been Clicked"
    End sub
    Sub Submit1_OnClick
      MsgBox "The Submit Button Has Been Clicked"
    End sub
</SCRIPT>
</BODY>
</HTML>
```

Advantages and Problems with Client Scripting

The primary advantage of client scripting is that the load is taken off the server to perform the processing.

There are two disadvantages to client scripting. The first is that the browser must support the script language and any extensions that you are using. Because of the variety of browsers, this means that you must write to the lowest common denominator if you are working on the Internet.

Surprises

The Web development group that I am associated with uses Microsoft tools and servers. Before we publish material on the World Wide Web, we test using several browsers, screen resolutions, and so on. Even when we think we understand what we are doing, we are surprised by the appearance of a page in a particular browser on occasion. The broad variety of browsers keeps Web development challenging.

The second disadvantage is that your creation is available for anyone to copy. When you see a page that you find particularly good, if the page is created using client scripting, a simple choice of **View**, **Source** will reveal all of the secrets.

Is the message that you should never use client scripting, and all scripting should be performed on the server using Active Server pages? Not quite. There are situations (particularly in an intranet environment, where the browsers are known and controllable) in which a combination of Active Server scripting and client scripting can be a functional combination. I find "always" and "never" to be words that should be used with caution.

Examples of Client Scripting

On the CD that comes with this book, there are two files, Default.HTM and Default2.HTM located in the Chapter 20 directory.

These are intentionally simple examples. In them you will see button controls and OnClick event handler script. To run these sample pages, all that you should need to do is double-click the file name in the Windows Explorer and the file will open in your default browser. If your default browser is Internet Explorer, the script will run (since it is VBScript).

The Least You Need to Know

You have examined some of the HTML intrinsic controls and the use of VBScript to use the events and properties of these controls to perform processing. All of the processing that was performed here was on the client browser.

- Although VBScript was used for the illustrations in this chapter, JScript could have been used.
- Scripting will always be a mixture of HTML and the script language. The objects that are being manipulated by the script language are usually HTML objects.
- Scripting is extremely easy to learn. Anything that you learn for the client scripting will usually apply for the Active Server Scripting.

Chapter 21

HTML Layout Control—Put It Where?

In This Chapter

- Create an HTML Layout Control
- Insert the HTML Layout Control into an HTML Page
- Create VBScript to activate the HTML Layout Control

One of the frustrations of working with HTML is that you cannot always precisely control the location of various objects in the Web page. Microsoft created a special ActiveX Control that deals with this issue, named the HTML Layout Control. A special file is created to contain the properties and settings of the HTML Layout Control which allows the same control to be used on multiple HTML pages. It is also possible to use multiple HTML Layout Controls on the same HTML page.

Visual InterDev incorporates tools to enhance the ease-of-use of the HTML Layout Control. An HTML Layout Control Wizard will create both the control and the HTML page at the same time.

Using the Template HTML Layout Wizard

The first step in demonstrating the HTML Layout Wizard is to create a Web project. Layout_Demo has been chosen as the name for the project. (A copy of this project is on the Companion CD.)

Creating Projects
If you encounter any problems in the process of creating a new project, refer to Chapter 6, "Creating and Editing Workspaces, Projects, and Files." A review of the procedures there should help.

Choose **File**, **New** and click the **Project** tab. Select the Web Project Wizard by highlighting it and enter the name **Layout_Demo**. Make sure that **Create new workspace** is highlighted and click the **OK** button. You will be asked on which Web server you wish to create the project. Enter the Web server name and click the **Next** button. On the next dialog box, select **Create a new Web** and click the **Finish** button. You will see the Workspace window opened with the Layout_Demo project open.

The project contains two files at this point, the global.asa and the search.htm that are created by default.

Using the Template Page Wizard

The Template Page Wizard performs the task of creating the layout control file (which has an extension of .ALX) and the accompanying HTML file at the same time. The process is started by choosing **File**, **New** and clicking the **File Wizards** tab. Select the Template Page Wizard by highlighting it as shown in the following figure.

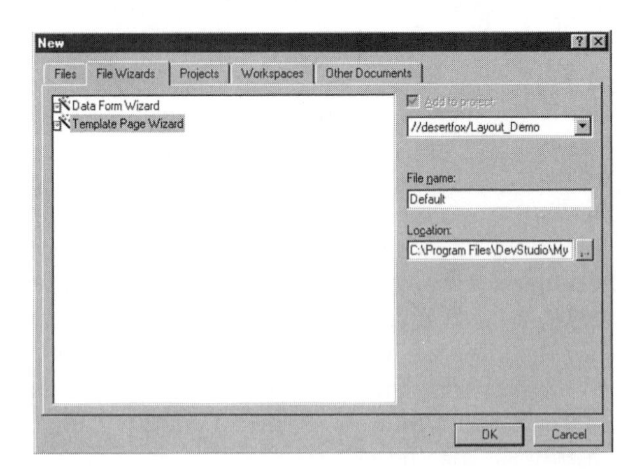

The name Default is used because this is the start page for the Web.

Enter the name **Default** in the File name text box and click **OK**. The next dialog box of the Template Page Wizard asks which template should be used. Select **LAYOUT** and click the **Finish** button as shown in the following figure.

Next, the Template Prompt dialog box asks for the title for the HTML page. "HTML Layout Control Demonstration" has been entered in the following figure.

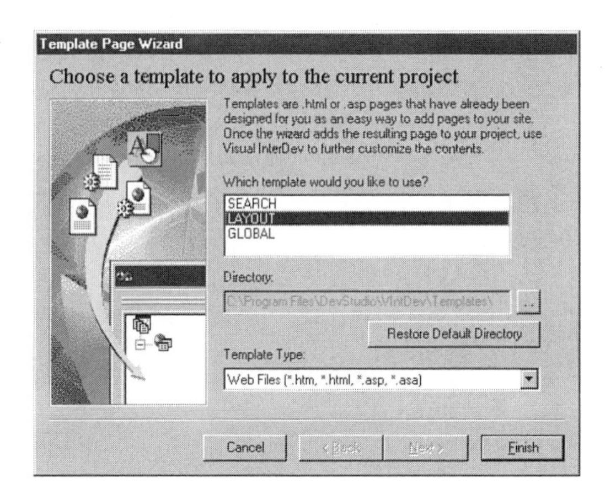

Global.ASA files and Search.HTM files can also be created with this Wizard.

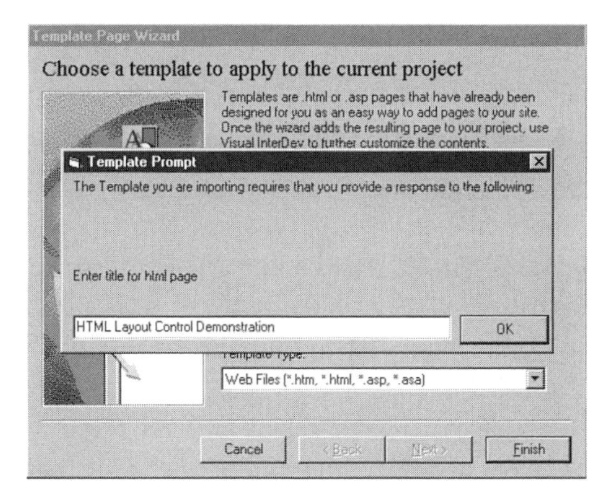

The title is shown in the titlebar of the browser.

The new page, Default.ALX is opened for editing. In the following figure, also notice that a file Default.HTM has been created. We also see the Toolbox for the Layout Control.

Customizing the Layout Environment

The HTML Layout Environment can be customized. Choose **Tools, Options** and click the **HTML** tab. The grid spacing and snap to grid can be set here.

The HTML Layout toolbar is also opened. This toolbar moves objects forward and backward on the control.

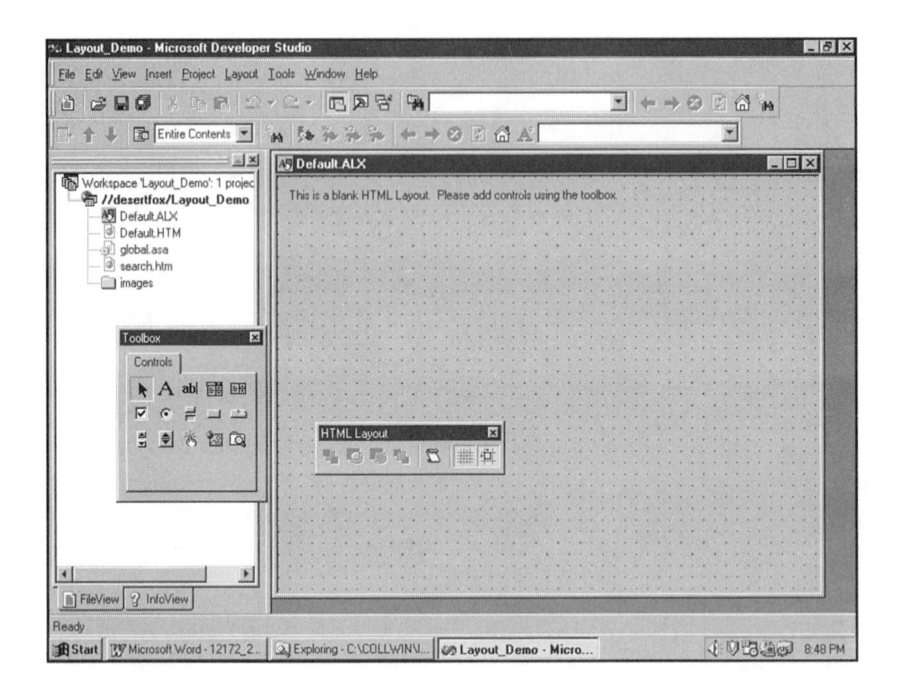

The Toolbox contains the controls that can be added to the Layout Control. You will recognize the controls listed below as standard Windows Controls. (Each of these controls has a set of properties and methods just as any Window Control.)

- **Label** Text label control.
- **Text box** Text box for input and display of text.
- **Combo box** Combination text and drop list box.
- **List box** Select list box.
- **Check box** Yes or No check box.
- **Option button** Radio button control. Usually a selection from a group of options.
- **Toggle button** Two state On/Off button.
- **Command button** Create a click event to initiate action.

- ➤ **Tab strip** Allows the selection of different forms.
- ➤ **Scroll bar** Can be set for either vertical of horizontal scroll. It is used with another control.
- ➤ **Spin control** Used to move through a percentage or number setting.
- ➤ **Hot spot** The mouse pointer can be changed over the hot spot and an action specified for a click event for instance. Useful as a map.
- ➤ **Image** Container for an image file.

Now add a Text box control by clicking the Text box on the Tool bar and draw the control on the HTML Layout as shown in the following figure. When you double-click the control, the Properties dialog box opens.

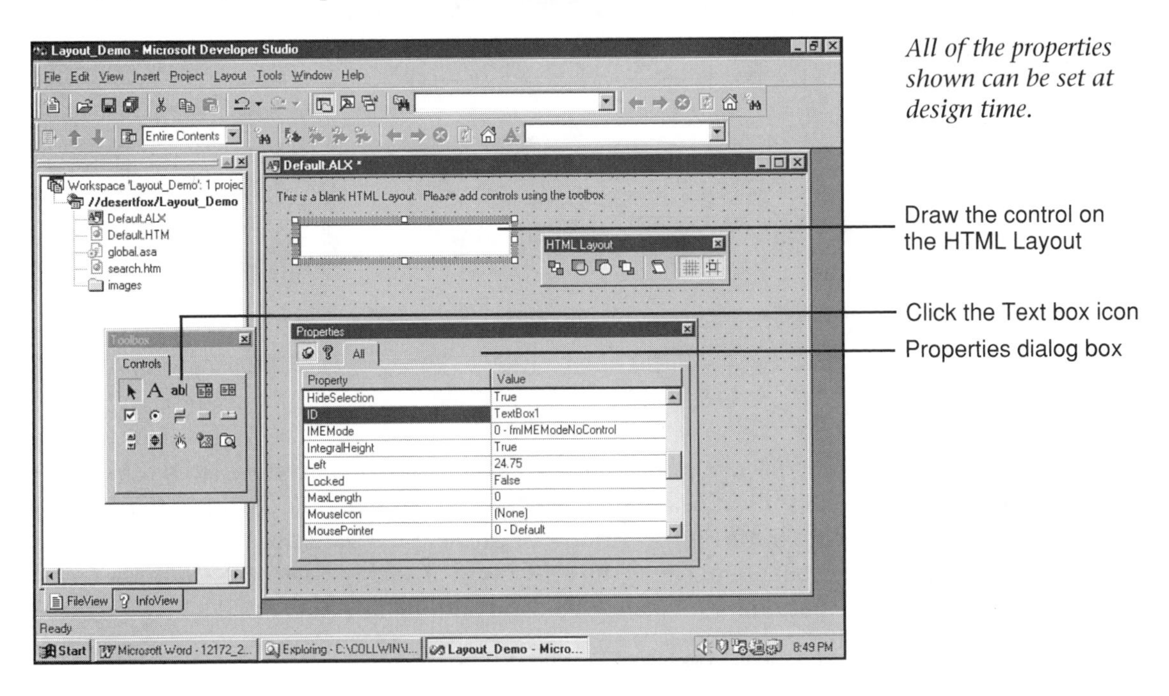

All of the properties shown can be set at design time.

The ID property is the name that is used to refer to the control. For this text box, the ID is TextBox1. (We will use this later when we are writing VBScript for the Layout Control.) The other property that we will use is the Text Property.

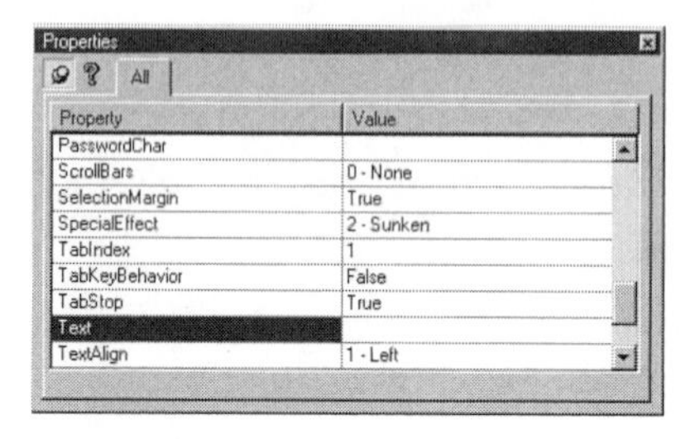

The design time values are shown in the properties in the Properties dialog box.

Looking at Default.HTM

It is now time to close the Default.ALX file. We could have added more controls. The only limit in the number of controls is set by performance and space in a Web page.

Earlier, we noted that a file named Default.HTM had been created. Double-click the file in the Workspace window and the file is open for editing.

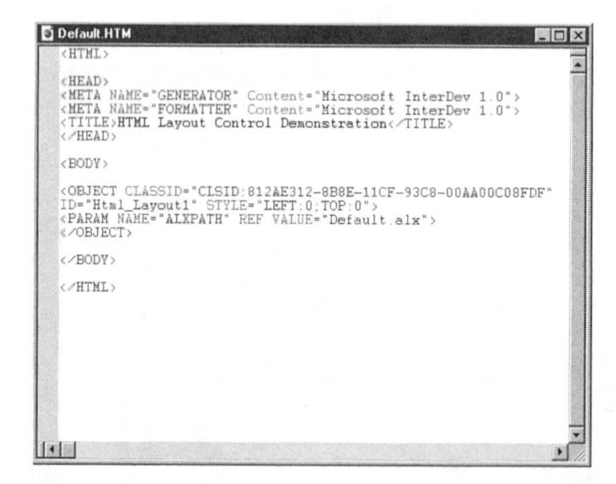

All of the standard elements and the HTML Layout Control were inserted into the file by the Wizard.

We can now add some VBScript to activate the Layout control that we inserted into the Web page. Place the cursor just below the <BODY> tag and right-click. Choose the **Script Wizard** from the menu. In the Event window of the Script Wizard dialog box, open the tree for the HTML_Layout1. OnLoad is the only event available. Select the **Code View** option button at the bottom of the dailog box and add the line of code

```
HTML_Layout1.TextBox1.Text = "It Works!!!"
```

as shown in the following figure.

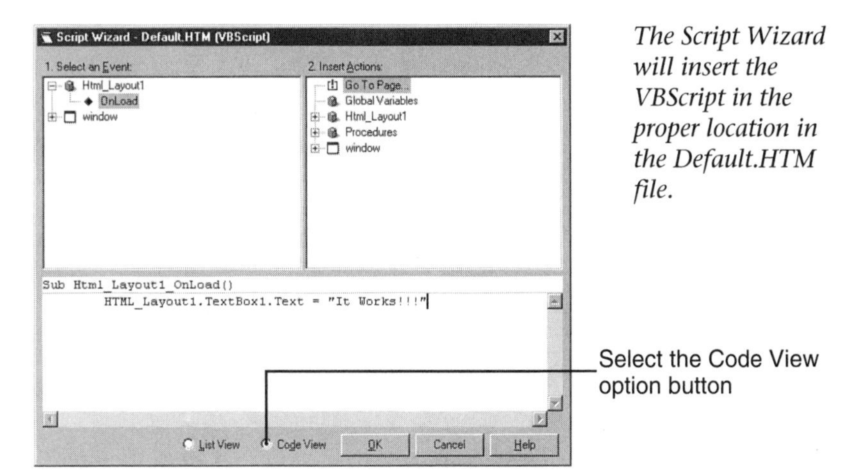

The Script Wizard will insert the VBScript in the proper location in the Default.HTM file.

Now click the **OK** button. You will see that the Script Wizard has taken care of all of the syntax required for insertion of the VBScript. In the figure below, you see the HTML in the Default.HTM file.

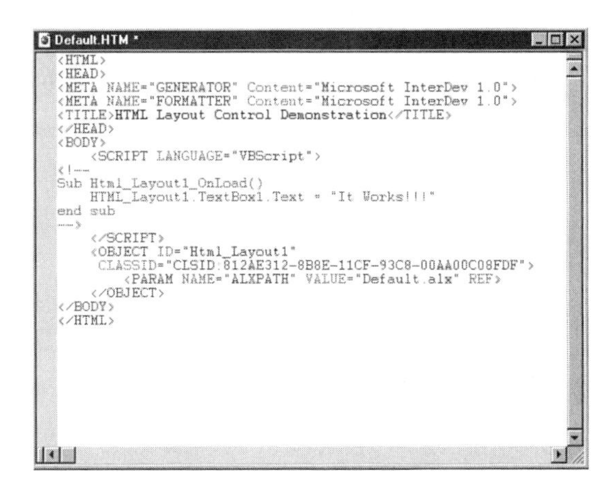

The Wizard also took care of inserting the Class ID (CLSID) for the HTML Layout Control. Before Wizards this was an error-prone task.

Testing the HTML Layout

It is now time to test our masterpiece. We accomplish this by right-clicking the edit window and choosing **Preview Default.HTM** from the menu. If we did it correctly, it should look like the following figure.

The file could also have been opened in the browser by setting the URL to http://servername/Layout_Demo.

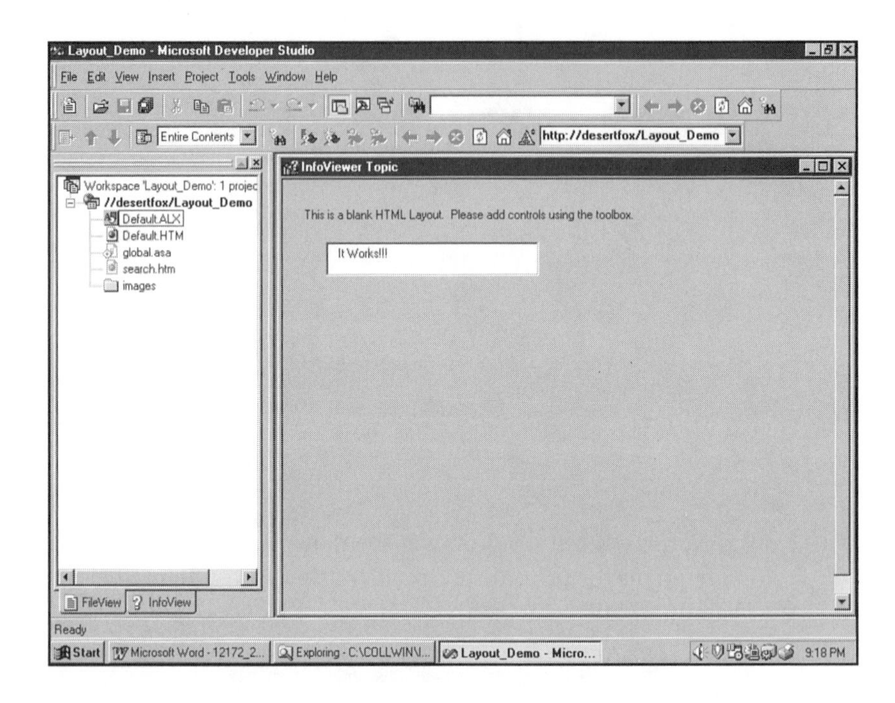

This demonstration has been very elementary for this reason: Placing ten controls on the HTML Layout Control would have only made the example harder to follow and would not have illustrated any more principles.

Creating an HTML Layout Control

You will not always want to use the Wizard to create an HTML Layout Control .ALX file. You may be creating a HTML Layout that will be used with several HTML pages. Creating an HTML Layout Control that is used on several pages gives a professional uniform look to your pages.

Choose **File**, **New** and click the **File** tab. Select the HTML Layout, enter a name and click **OK**. The HTML Layout edit window will be opened with the new form as it was above. You follow the same procedure as before for adding controls. The difference is that an HTML page will not be automatically created as with the Wizard. You will need to insert the HTML Layout Control into the HTML Page.

To perform the process of insertion in the HTML page, simply open the page in an edit window, place the cursor and right-click between the <BODY> tags. Choose **Insert HTML Layout** from the menu and select the HTML Layout to be inserted.

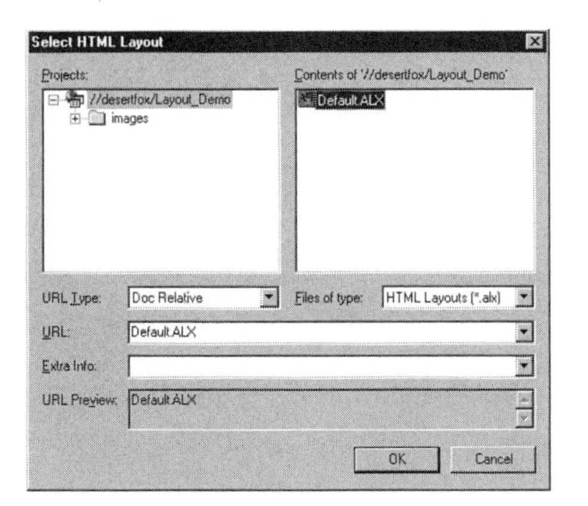

The same .ALX file can be used in many different HTML documents.

When you click the **OK** button, you will see that the HTML Layout Control has been inserted in the HTML file.

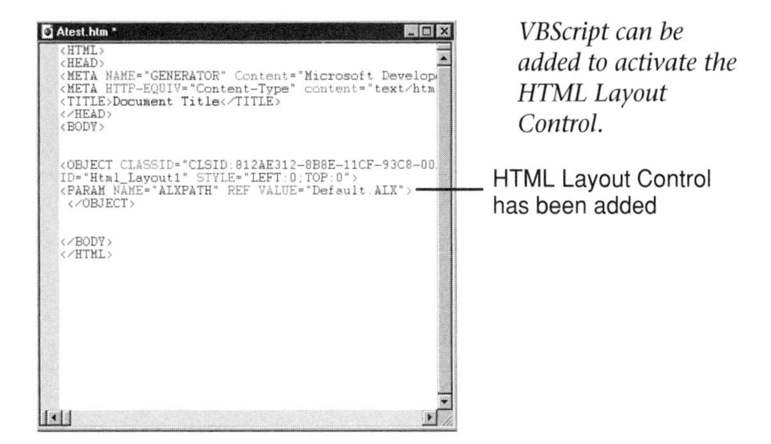

VBScript can be added to activate the HTML Layout Control.

The Least You Need to Know

The HTML Layout Control is a very useful and powerful ActiveX Control. It can save you many hours of tedious coding and arranging. Simply create the layout that you want one time and use it over and over on different Web pages and projects. Just remember:

- Use the Wizard to create the HTML Layout Control. It will perform the task properly.
- Use the facilities of Visual InterDev to insert the HTML Layout Control into a Web page. The CLSID will always be correct.
- Use the Script Wizard to generate the correct framework for the VBScript. It will save you from many mistakes.

Chapter 22

Internet Explorer Script Debugger: It did What?

In This Chapter

- Learn what the Internet Explorer Script Debugger can do for you
- Learn where to get your copy of Internet Explorer Script Debugger
- Learn why you may not want to use Internet Explorer Script Debugger while it is in Beta release
- Learn how to install and uninstall Internet Explorer Script Debugger

Visual InterDev is missing one feature. (You knew there had to be at least one.) The feature that it is missing is a script debugger. That is why Microsoft Internet Explorer Script Debugger is so useful. It is in Beta release, and is available as a free download. The file name is msie3dbg.exe and it is available at **http://www.microsoft.com/msdownload/scriptie.htm**. The price is right, but there is a risk with Beta software. (It was in Beta release at the time of writing. I hope it is in final release as you read this.)

It is designed to run with Microsoft Internet Explorer 3.01 Build 1215. This means that it also runs as part of the browser in Visual InterDev.

What Beta Release Software Really Means

Before you rush to the Microsoft download site and get this marvel, make sure that you have evaluated the risk. Beta release software means two things. The first is that you are warned that the software may mess up your system to the point that you have to delete everything on your hard drive and start over. I have done it. It is not a pretty process. If this happens and you go to the software vendor for help you learn the second thing, they may reply that after all it is Beta and you know the risk. Terribly sorry and all that, but tough luck. The reason that vendors do Beta releases of software is to get people to find what is wrong with the software. You are essentially a Beta Tester. Read the EULA (End User License Agreement) from Microsoft on Beta Software, specifically clauses 8, 9, and 10. Trust me, there will be bugs. Whether the problems are destructive or not is the risk that you are taking. The question now comes down to how great is the risk of installing Beta Software from Microsoft. My experience is that it is not a great risk. But if it eats your system, I am sorry but you know the risks of Beta software.

Installing and Uninstalling Internet Explorer Script Debugger

Once you have obtained your copy of Internet Explorer Script Debugger, you will need to install it. Simply place the file that you downloaded in any temporary directory and double-click it in the Windows Explorer. It is a self-extracting, self-installing file. It finds its way and installs with nothing more required than your consent to the EULA (End User License Agreement).

Even Though It IS Beta

The IE Script Debugger appears to be a complete product at this point. I use it regularly and have not found a lot of "I wish it did this" issues.

If you decide that you want to uninstall the Internet Explorer Script Debugger, choose **Start**, **Settings**, **Control Panel**. Double-click the **Add/Remove Programs** icon. Click the **Install/Uninstall** tab. Find the program named Microsoft Script Toolkit for Internet Explorer and click the **Add/Remove** button. The Internet Explorer Script Debugger is now gone.

Script Debugger Works for VBScript, JavaScript and JScript

Script Debugger will work for both VBScript and JScript. Because of the design of Script Debugger, it should work for other script languages as well.

Starting Internet Explorer Script Debugger

You can start the Script Debugger in four ways. In every case, the Internet Explorer or the Visual InterDev browser window must be running for the Script Debugger to run.

- Choose **View**, **Source** in the Internet Explorer. The source file is displayed in the Script Debugger. If you do not have the Script Debugger installed, the source file is displayed in Notepad.
- In Visual InterDev, when you are previewing a page in a browser InfoViewer Topic window, right-click the InfoViewer Topic window and choose **View Source** from the menu and the source file will be displayed in the Script Debugger.
- Choose **Edit**, **Break at Next Statement** in Internet Explorer or Choose **Debug**, **Break at Next Statement** in the Script Debugger.
- Execute a Stop statement in a VBScript or a debugger statement in a JScript.

Internet Explorer Script Debugger Windows

Internet Explorer Script Debugger has five types of windows. Each provides a different set of information.

Edit Window

The Edit Window opens when Script Debugger is started. As you see in the figure below, it is the parent window for Script Debugger, which is an MDI (Multiple Document Interface like Microsoft Word where multiple documents can be open) application.

The **View** menu is used to open or select the Project Explorer, Immediate, and Call Stack windows. The **Debug** menu provides access to the debugging commands, which you can use to single-step through a script.

The Internet Explorer Debugger parent window acts as a container for the other child windows where the work is done.

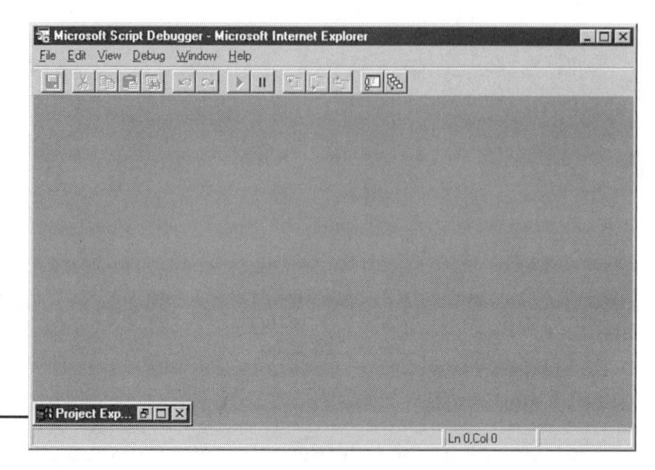

The Project Explorer child window can be minimized but not closed.

Project Explorer

This window provides a hierarchical list of all currently open HTML files as shown in the figure below. When a file name is double clicked in the Project Explorer Window, the file is opened in a Code Window.

There will be only one Project Explorer Window in a Script Debugger session.

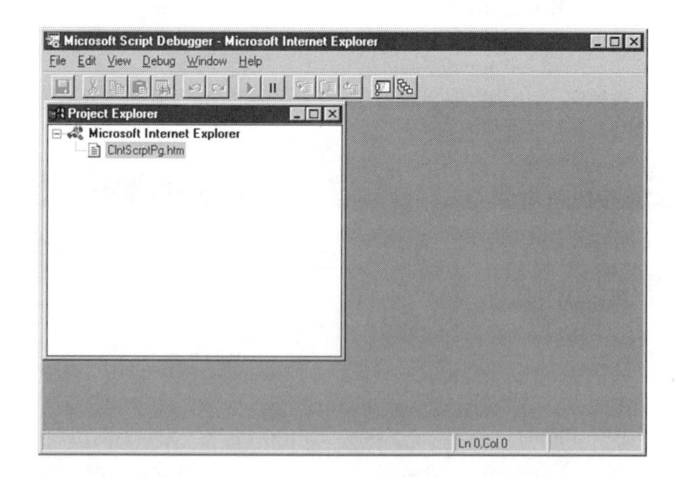

Code Window

The Code Window is where the HTML and script code are displayed for debugging and editing as shown in the figure below.

Breakpoints can be set and cleared on script statements using the Debug menu. When you attempt to change the contents of the code displayed, Script Debugger will inform you that the original file is stored on an HTTP server and offers to open a local copy for editing. If you edit the local copy, any changes will need to be made to the original copy at a later time. This is an excellent safety feature that allows you to try changes without fear of damaging or replacing the original file.

You can have multiple Code Windows open during a debugging operation.

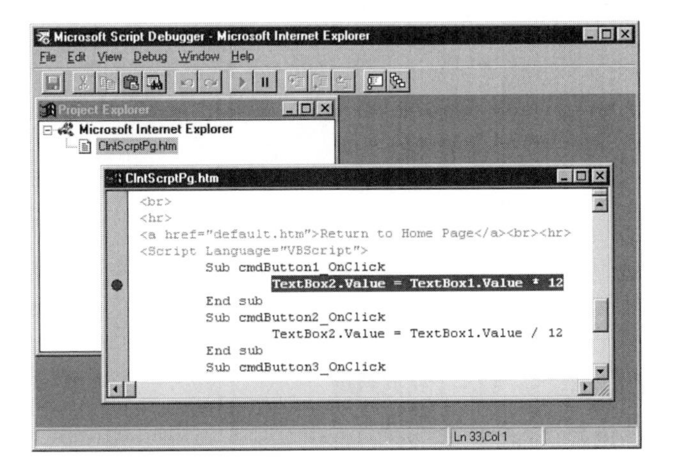

A breakpoint has been set on a VBScript statement. Breakpoints are allowed only on script statements.

Immediate Window

The Immediate window is used to display information during the execution of a script as shown in the figure below. The Immediate window is only active when the script is at a breakpoint.

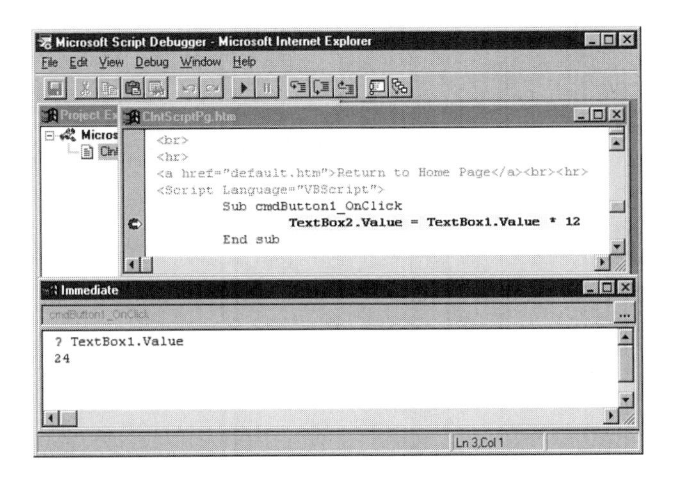

The value of a variable is examined at the breakpoint.

Call Stack Window

The Call Stack Window displays a list of all currently active procedures as shown in the following figure.

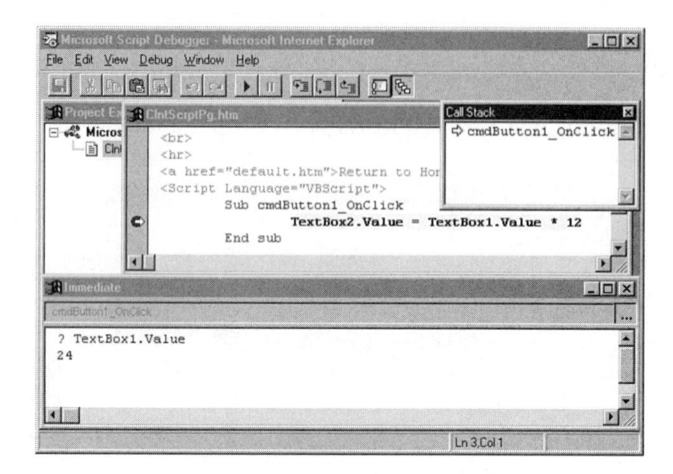

The arrow indicates that the cmdButton1_OnClick procedure is currently active.

The Least You Need to Know

Internet Explorer Script Debugger is another tool to use in your quest to create functionally flawless Web content. It is especially useful in determining where a problem is located. It can be of great assistance in allowing you to single-step through a procedure, testing variable contents at each step. When a procedure runs but produces the wrong answer, locating the error can be maddening if you can't see each step of a calculation. The Script Debugger lends a helping hand. Just remember:

- ➤ While Internet Explorer Script Debugger is in Beta release, there is a risk to installing it on your system. Just because I have it installed and haven't encountered any problems, doesn't mean that you won't.
- ➤ Experiment with the use of Script Debugger. Because you can only make changes to local copies of the file, you can't hurt anything by experimenting.

Part 6
Finishing Touches

Web pages are not the same old boring stuff any more. The Web is becoming multi-media. It is becoming very common to surf into a Web site and hear music. Pictures are becoming more sophisticated.

Microsoft Music Producer to the rescue! Now you can create your own music to put on your Web pages royalty free. Microsoft Image Composer to the rescue! Now you can use art that is royalty free. (And the Microsoft Media Manager is there to help organize your creativity.)

Chapter 23

Arts and Crafts: Media Manager, Image Composer, and Music Producer

In This Chapter

- Add Image Composer and Music Producer to your Visual InterDev Tools menu
- Create images with Image Composer
- Create music with Music Producer
- Store and index your multimedia files with Media Manager

When you have no musical or artistic ability, applications such as Music Producer and Image Composer seem to be magic. Do not avoid this chapter because you think you have no artistic bent—it is impossible to have less artistic or musical talent than I have (even my dog can out-sing me, and people yell at him to keep quiet). Image Composer and Music Producer make it possible for anyone to create art and music for their Web site.

Check This Out...

Royalties and Copyright Infringement

There is one very serious consideration when using art or music on a Web site. Because you are publishing the work, it is easy to infringe upon someone else's copyright. Image Composer and Music Producer take away that worry. All music created by Music Producer is original and the copyright is yours. (You probably won't win a Grammy, but it is yours.) All images that come with Image Composer are provided royalty-free, so you have no worries.

Including Image Composer and Music Producer in the Tools Menu

The first task is to add the Image Composer and Music Producer to your Tools menu in Visual InterDev. If you have forgotten how this is accomplished, refer to Chapter 5, "Toolbars/Menus/Help and Documentation: What Does This Do?" in the section titled "Menus." The process is explained there. If you choose not to add these to your Visual InterDev Tools menu, you can always open these programs from the Start menu list.

Image Composer

Image Composer has a myriad of features and functions: You can compose images of many different separate pictures, apply effects to the pictures, and generally perform more changes than we have room in this book to document.

Let's look at an example. In the following figure are two pictures. The first is just as it was loaded from the photos that come with Image Composer. The second is shown after applying the Fresco Art Effect (Fresco paints in a coarse style using short, rounded dabs). The combinations are almost limitless.

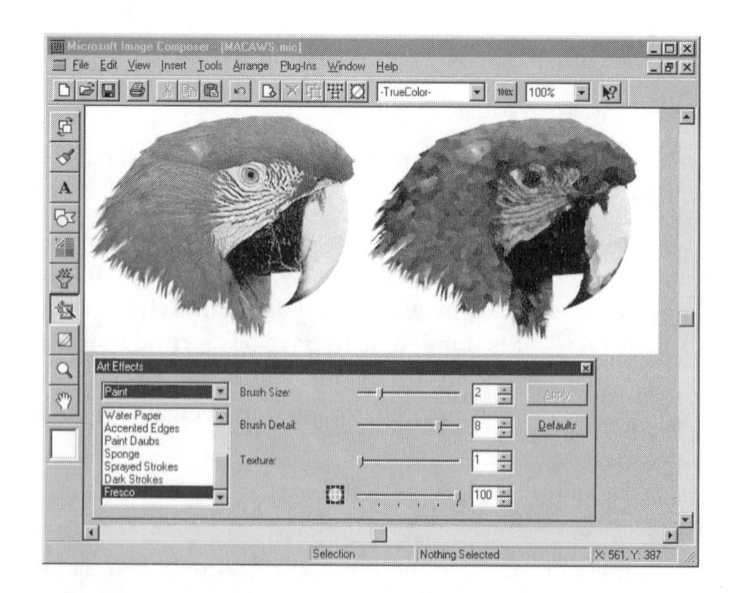

The image clarity is dependent upon the capabilities of the monitor and the color settings. This monitor is set for 256 colors at 600×800 resolution.

Image File Formats

There are many different image file formats. The native format for Image Composer is .MIC (Microsoft Image Composer Format). There are also two formats that are commonly used on the Internet and supported by Web browsers—.GIF and .JPG.

The advantage of .GIF and .JPG format files is that they are reasonably compact. To illustrate this point, when a picture was saved in the same format, the size varied as follows:

- .BMP (Windows Bitmap) 573KB
- .MIC (Microsoft Image Composer) 336KB
- .GIF (CompuServe GIF) 40KB

It is easy to see why .GIF is a good format for the Internet. Every image file in your Web page has to be transferred over a connection (in many cases a slow connection). At 14.4KB, a 4K file takes 2.2 seconds. The files in the illustration would require 22 seconds for the .GIF and 5.3 minutes for the .BMP file. (Be aware that in less than 5 minutes the visitor to your Web site has left in disgust.)

Which File Type Do I Use?

The two file types that are used most often and most widely supported are .GIF and .JPG. There are times when one will be preferred over the other. As an example, .JPG is the preferred file type for photos that use the entire color spectrum. If you are going to understand the ins and outs of graphics, file types, and compression methods, you have a large study ahead.

As you are looking for image files that you can use, it is useful to know that Image Composer reads and writes image files in the following formats:

- .TIF TIFF
- .GIF CompuServe GIF
- .TGA Targa
- .JPG JPEG
- .BMP Windows Bitmap
- .PSD Adobe Photoshop

Understanding Sprites

When you are working with Image Composer files, you need to understand a sprite. An Image Composer image is made up of two parts, the *sprite,* which is the "picture," and the transparent background. The *sprite* can be moved from the background. Multiple sprites can be stacked one on top of another. The order of the pile can be changed. This is similar to other graphics programs that work with layers but is far easier to manipulate.

Here, the sprite has been moved from the background. There are handles around the sprite that are used for manipulating the image.

In the following figure, another sprite has been added to the image. If the image were to be saved at this point, only the portion of the sprites that are located on the white background area would form part of the image.

Sprites are stacked on top of each other. The order can be changed.

MIC versus .GIF or .JPG

If the image is saved in the .MIC format, the identity of the separate identity of the sprites is preserved as separate objects. If the file is saved as a .GIF (as you would when placing it in a Web page), the sprites are merged into one image and the sprites can no longer be manipulated separately.

A Palette for Every Taste?

The Tools menu in Image Composer can be used to open eight different Tool palettes (see the following figure for an example of the Arrange palette).

- **Arrange** Used to arrange sprites
- **Paint** Used to color images
- **Test** Used to add text to an image
- **Shapes** Used to add geometric shapes to an image
- **Patterns and Fills** Used to fill areas with patterns and colors
- **Warps and Filters** Used to apply distortions to images
- **Art Effects** Used to make a photo look like a painting
- **Color Tuning** Used to perform color changes and filtering

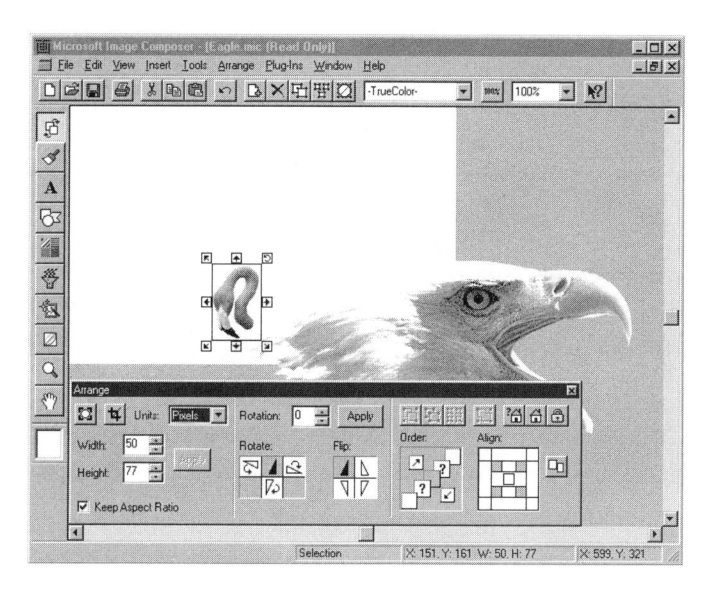

Because the Keep Aspect Ratio check box is checked, the proportions of the sprite remain constant during the resize operation.

Note the following capabilites of the Arrange palette (remember to experiment with this and the other palettes to discover all their capabilities):

- **Rotate** This rotates the image in 90 degree moves, right or left.
- **Flip** This mirrors the image horizontally or vertically.
- **Order** This changes the order of a sprite in a stack of sprites.
- **Align** This adjusts the alignment of one sprite relative to another sprite.

Learn More in the Help Tutorial

The Help files contain an excellent tutorial on the Image Composer that is time well spent. Your tour here has only scratched the surface of the capability of Image Composer. (I concentrated on those aspects that are not obvious, such as the concept of the sprite. Once you have mastered the manipulation of sprites, you are well on your way to mastering Image Composer.)

Finally, remember that Image Composer comes with 200+MB of photo images to start with. Image Composer makes even an artistic klutz such as me begin to feel like a Michaelangelo or Van Gogh. (I haven't cut off my ear yet. I wonder if that would help?)

Music Producer

The Music Producer is very easy to use. There are six basic aspects of the music you will create that you need to set, and the method that you use to set these characteristics is very simple. You click the **Preview** button and begin changing until it is what you want.

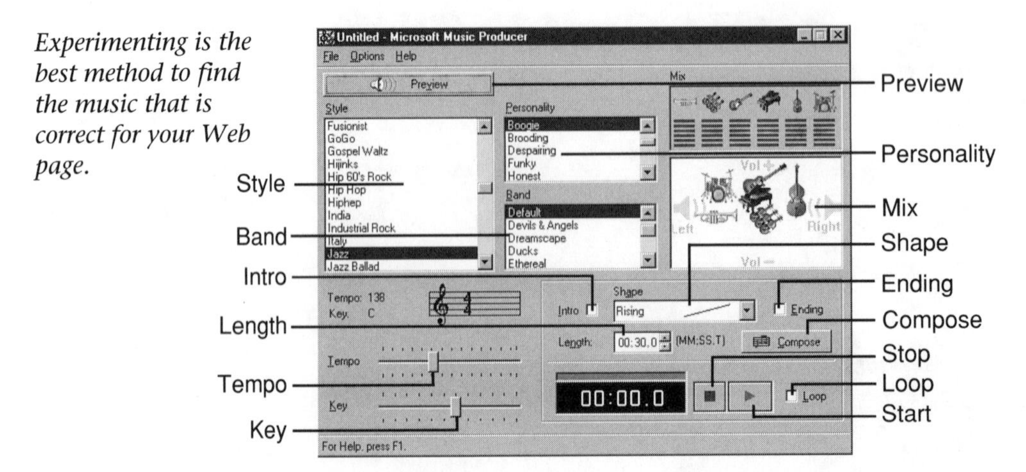

Experimenting is the best method to find the music that is correct for your Web page.

Changing the Sound

The Style is the factor that changes the music the most. The important factor is that you can press the Preview button and while the music is playing, you can change any of the properties that change the sound of the music. As an example, Jazz is the default style setting for the first time that you open Music Producer. To change to another Style, just click the name. Use the same method for the Personality and Band. Tempo and Key are slide bars. Mix is the one that is not immediately obvious. Clicking one of the instruments and dragging it to another area of the Mix window will increase or decrease the relative volume of the instrument. Moving the instrument right or left will shift the speaker balance. You can see an example of this in the previous figure.

Composing Music

When you are ready to create a musical recording, click the **Compose** button and a new unique recording will be created. The settings for a composition are:

- **Intro** This creates a beginning flourish.
- **Shape** A rising shape will cause the music to rise in volume and presence.
- **Ending** An ending resolution is created.
- **Length** The length of the composition is set in seconds.
- **Loop** If the music is to play in a continuos loop, this will allow you to hear the loop transition.

After you have adjusted the settings, click the **Start** button and sit back and listen. If you like what you hear, you can save the file.

Composing

Even though you click the Compose button to create your musical work, the computer is the one composing. You are mixing and matching general characteristics of the music. It is somewhat similar to the rhythm section on an electric organ or electronic keyboard.

Music File Formats

The native format for Music Producer is the .MMP format. This format is not recognized by most Web browsers. The reason that you will save music files in the .MMP format is so that they can be opened and edited using Music Producer. Music Producer has no editing capability in this release, so it is a matter of being able to open a file and listen to it using Music Producer.

Another format supported by Music Producer that is also recognized by most Web browsers is the .MID format. The other popular format is the .WAV format. The advantage of .MID files is that they are much smaller than the equivalent .WAV file. In fact, one test indicates that the .MID format is one-tenth the size of a .WAV file. Where every byte must be transferred, size is important. A file can be saved in the .MID format but cannot be opened in this format. When you are satisfied with a composition, save it in the .MID format for use on the Web.

Media Manager

Media Manager is useful for organizing multimedia files. When you first install Media Manager, the introduction window shown in the figure below will appear. There is a check box that will prevent this startup screen from appearing, but I suggest that you let it appear until you have become familiar with Media Manager.

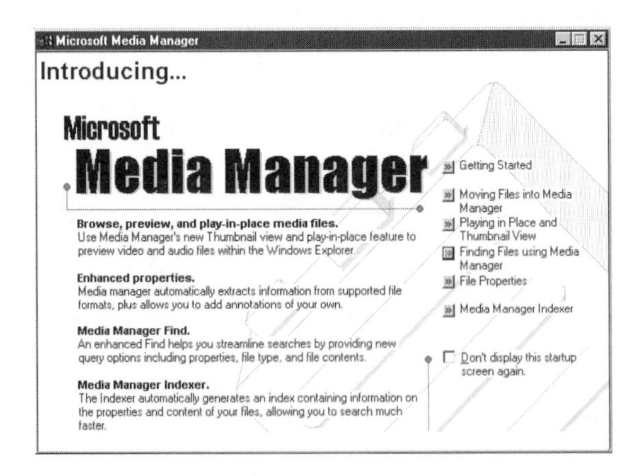

The Media Manager startup screen is an easy path to the Help files.

When you first install Media Manager, you will be asked if you want to convert the Multimedia Files directory to a Media Manager directory. If you say yes and later change your mind, you can convert it back to a standard directory. One caution before you convert: If you remove Media Manager from your system, the Media Manager directory will be unusable. After the conversion, you will notice two different things:

➤ **New Folder** The folder has a new icon.

➤ **Thumbnail** When you click the Thumbnail icon (now added to the toolbar), you will see thumbnail images of the picture files in the directory. This can make finding a picture much easier (and may be proof that a picture is worth a thousand words).

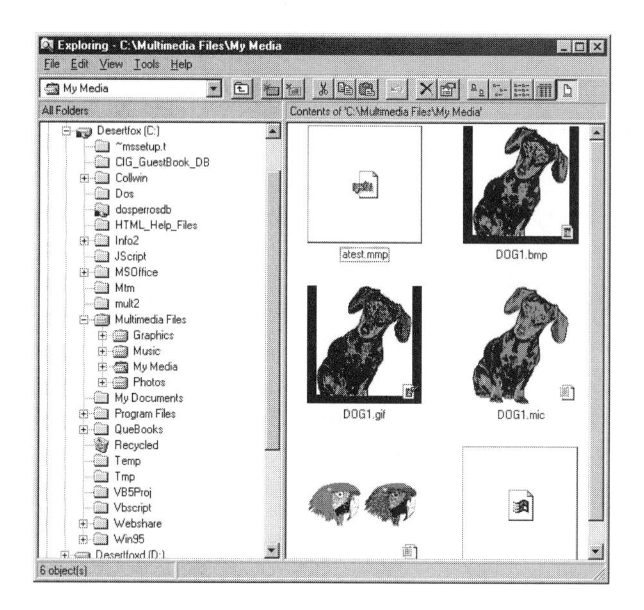

The thumbnail images can be of all supported image formats.

There is additional annotation information available on the files in the Media Manager directory. When you right-click a file name and choose Properties from the menu, you will have a dialog box as shown in the figure below. You can add data in many categories. These categories can be searched to find a file. When you want to search for files based on the annotations, in the Windows Explorer, choose **Tools**, **Find**, **Files**, or **Folders in Media Manager**. This allows searching based on the Media Manager Annotations Database.

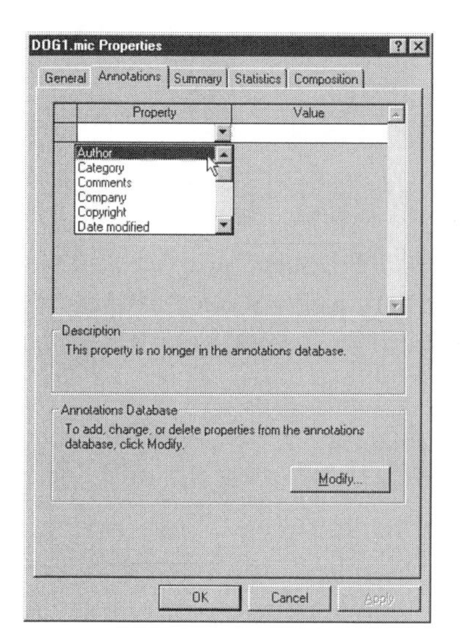

Click the Modify button to change any existing values in the Annotations Database.

- **General** Provides location, creation, modification, and last access date information.
- **Annotations** Provides the ability to add annotation information to a file.
- **Summary** Summary information about a file.
- **Statistics** Statistical information about a file.
- **Composition** Provides color and size information for image files.

Where Did My Help Go?

After a while, you will be familiar with Media Manager and you will turn off the startup screen. This is when you realize that you can no longer find the Media Manager Help file. It is a file named MM.HLP and it is hiding in your Windows\System directory on Windows 95 and the WINNT\System32 directory on Windows NT. When you double-click it, you will see the Help for Media Manager screen shown in the following figure.

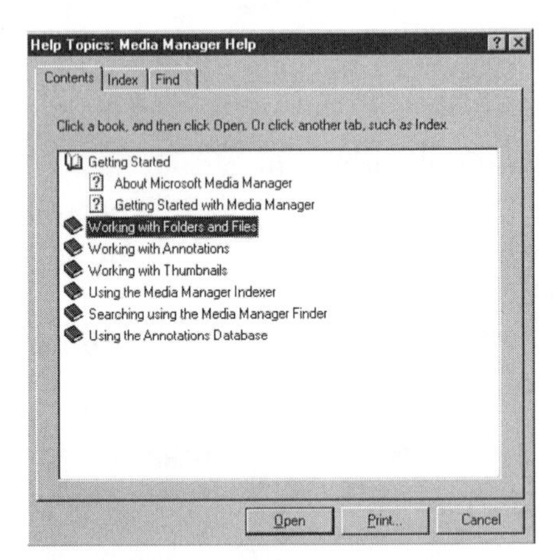

This Help file will answer most of your questions about Media Manager.

The Least You Need to Know

Image Composer, Music Producer, and Media Manager are very powerful tools. Spending time to learn to use them well will pay large dividends. Finding royalty-free images and music can be a long and frustrating search. These tools will eliminate the hunt. Media Manager will keep your Multimedia files organized and easy to locate. Just remember:

- Convert your Multimedia Files directory to a Media Manager directory. Add annotation information as you create the files. It will work for you later.
- Explore the tutorial for Image Composer. It is excellent. There are so many aspects to the creation of images that you will never quit learning. You don't have to know it all to be creative, just be persistent and willing to experiment.
- You will never win a Grammy with Music Producer, but neither will you ever pay a copyright royalty. Experiment until it sounds right to you. Get the opinion of others. Music is art, not science.

Chapter 24

Creating Your First Sample Project: Where Are the Blueprints?

In This Chapter

- ➤ Create your own sample project
- ➤ Place the project on your Web site for use

In this chapter, you are going to create a sample project. I will be creating one also as a guide. The project that I will create will be on the companion CD for your reference. Borrow pieces from it and use it for whatever purpose you need. (The project that you create should be customized to your taste and purposes.) I hope that you will elaborate on yours more than I do on mine. I will keep mine intentionally simple so that you will find it easy to locate and extract any parts that you may find useful.

Check out the CD-ROM!

This chapter has corresponding sample files on the included CD-ROM. Simply click on "Examples," then the corresponding chapter number for which you are interested. (Be sure to read the Readme text file, also located in the "Examples" Section, for important information on installing the sample files.)

Project Specification

The purpose of my project is not to provide useful Web content or a useful application. The purpose is to employ different types of functionality and to link from one type to another. Viewed objectively, this project accomplishes nothing. It is, well, frankly useless, but sort of fun to create.

In this project you will find various elements that are discussed in other chapters. (We won't explain everything in detail here.) The chapters that you should refer to for further information are:

- Chapter 6, "Creating and Editing Workspaces, Projects, and Files." This covers the project and file creation functions.
- Chapter 7, "Active Server Pages—Don't Just Sit There!" The functioning of the Active Server is covered.
- Chapter 8, "VBScript: Lights, Camera, Action!" Review this material for more on VBScript syntax and functions.
- Chapter 12, "Relational Databases and SQL: Information, Please!" You will find guidance for the syntax of the database query here.
- Chapter 15, "Database Connectivity—Long Distance, Please." You will find the explanation for the use of ODBC here.
- Chapter 20, "Client Scripting—No, You Do It." The use of VBScript on the client side is covered here.
- Chapter 23, "Arts and Crafts: Media Manager, Image Composer, and Music Producer." The creation of the music and images is explained here.

You will see each piece as it is put into place, examine all listings to see how everything works, and learn a few tricks along the way.

Enough introduction. On with the show!!

Creating the Project

The first task in creating the project is accomplished by choosing **File**, **New** and clicking the **Projects** tab. Next, choose the Web Project Wizard by highlighting it. Enter the name for your project in place of CIG_SampleWeb as shown in the following figure, then click **OK**.

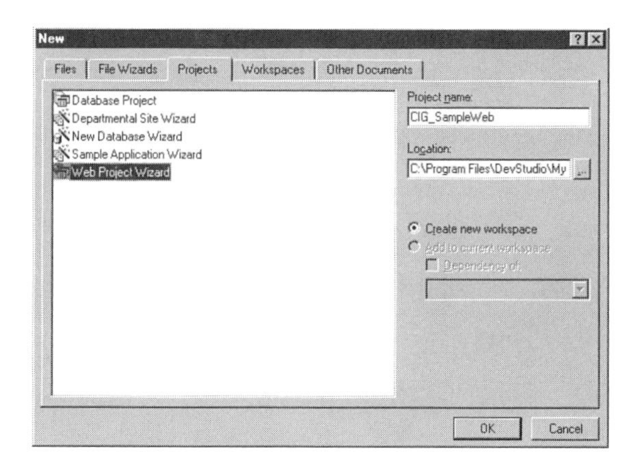

Make sure that you create this project in a new workspace.

In the Web Project Wizard (Step 1 of 2), you will enter the Server Name and click the **Next** button. In Step 2 of 2, make sure that **Create a new Web** is selected. If you want a search page, check the **Enable full text searching for pages in this Web** check box. Finally, enter the Web name and click the **Finish** button.

You now have the basic Web project created. The Web site is set up on your server with the necessary permissions.

Adding the Default.HTM Page

You need a place for visitors to start your Web. The usual is to create a Default.HTM file. When visitors come to your site with a URL and no file name, the Default.HTM file is dispatched to the visitor's system.

To create the file Default.HTM, choose **File**, **New** and click the **Files** tab. Select **HTML Page** by highlighting it, enter the name **Default.htm** in the **File name** text box and ensure that the **Add to project** box is checked.

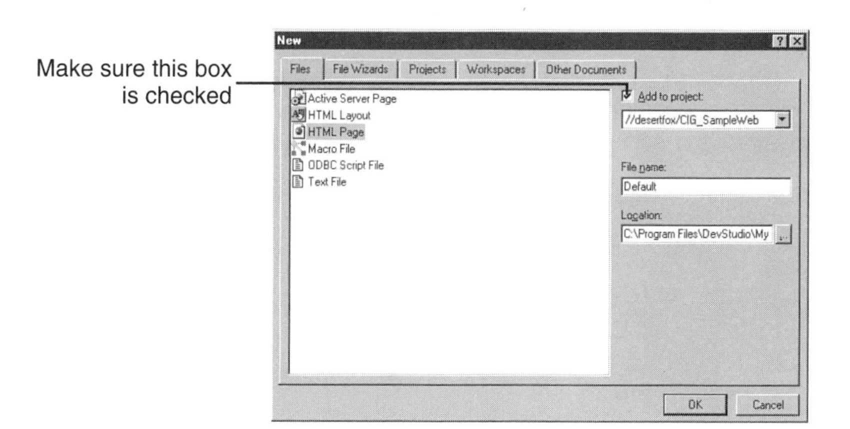

When creating a new file, you will usually want to add it to the current project.

Click the **OK** button. You now have a Default.HTM file that won't do much yet.

The design of your Default.HTM page is very important. It is the first impression that visitors will have of your Web site. (When you look at the Default.HTM page that I have created in the figure below, you will instantly realize that it is not intended to be a serious Web page design.)

Our founder, being a typical boss, sits on his tail and barks all day.

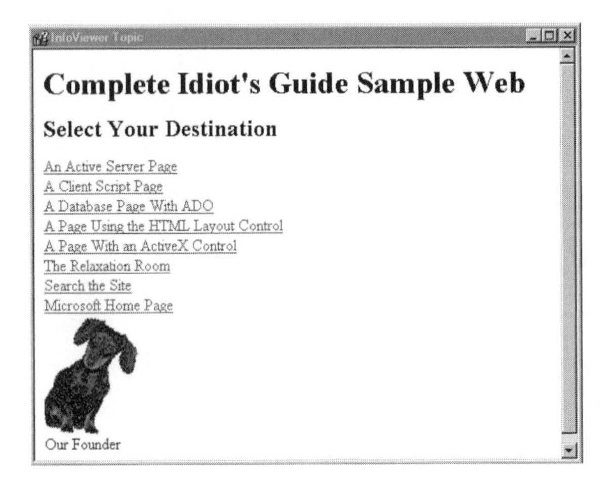

Examining the HTML in Default.HTM

Now let's look at the HTML in Default.HTM and see what was intended with each part.

These lines are generated by Visual InterDev (no change was made to them).

```
<html>
<head>
<meta name="GENERATOR" content="Microsoft Developer Studio">
<meta http-equiv="Content-Type"
content="text/html; charset=iso-8859-1">
```

The title "Complete Idiot's Guide Sample Web" for the page was entered in the next line. This is displayed in the title bar of the browser.

```
<title>Complete Idiot's Guide Sample Web</title>
</head>
```

Next, the background color is set to white rather than the default gray, then some music is added to create a mood. The music was generated using the Microsoft Music Producer, as was all of the music on this site. If you have a sound card and hear the music, you will know why I actually let others design the music for my Web pages. (This is fun for me, remember.)

```
<body bgcolor="#FFFFFF">
<bgsound src="images/logo.mid">
<h1>Complete Idiot's Guide Sample Web</h1>
<h2>Select Your Destination</h2>
```

These lines are the links to the other pages in the site and an external link to the Microsoft home page. This page serves as a gateway to the remainder of the site.

```
<a href="SampleASP.HTM">An Active Server Page</a><br>
<a href="ClntScrptPg.HTM">A Client Script Page</a><br>
<a href="DB_ADO_Page.ASP">A Database Page With ADO</a><br>
<a href="LayoutPage.HTM">A Page Using the HTML Layout Control</a><br>
<a href="ActiveXPage.HTM">A Page With an ActiveX Control</a><br>
<a href="RelaxPage.HTM">The Relaxation Room</a><br>
<a href="Search.HTM">Search the Site</a><br>
<a href="http://www.microsoft.com">Microsoft Home Page</a><br>
```

These lines insert the picture of "Our Founder" and adds the caption. This image was taken from the Microsoft Image Composer photo files. There is a wealth of material available there.

```
<img src="images/dog1.gif"><br>
Our Founder
</body>
</html>
```

And there you have it. In all, this page took about 30 minutes to create including the creation of the music and the creation of the image. I know that you are saying that you can do better than this. I agree, and your challenge is to do so. Good luck.

Adding an Active Server Page

The Active Server Page actually consists of two pages. The first is a form that is used to collect information from the visitor. This is an HTML form (SampleASP.HTM on the CD) as shown in the following figure.

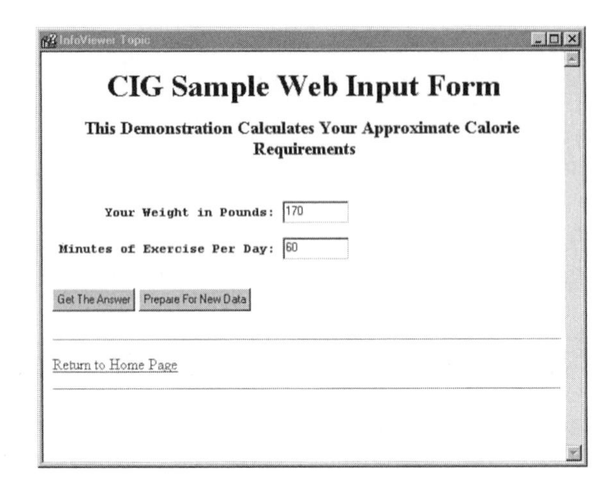

After the requested data is entered, the button (which is a Submit button) titled Get The Answer is clicked.

The HTML form is all standard HTML and is supported by most browsers. After the information is submitted, the server generates the answer using an Active Server page (SampleASP.ASP on the CD) and sends the answer to the client as shown in the figure below.

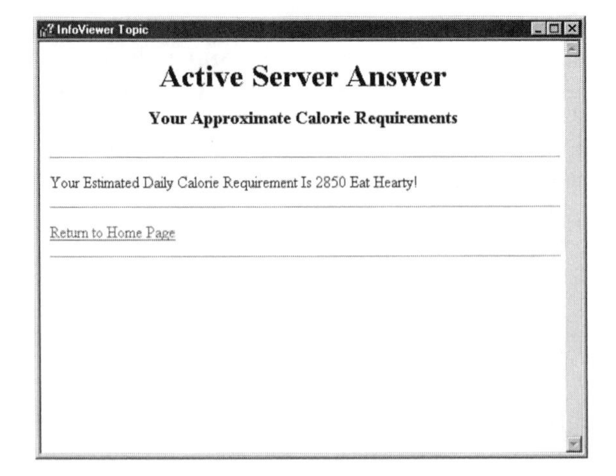

This HTML page was created by the Active Server reading the instructions on the Active Server page and generating standard HTML.

More HTML Code

Now it is time for a look at the HTML in these two pages. First let's examine the HTML form. This first section we understand from the Default.HTM page:

```
<HTML>
<HEAD>
<META NAME="GENERATOR" Content="Microsoft Developer Studio">
```

```
<META HTTP-EQUIV="Content-Type" content="text/html; charset=iso-8859-1">
<TITLE>Sample Web Active Server Input Form</TITLE>
</HEAD>
<BODY BGCOLOR = "#FFFFFF">
<bgsound src="images/sampleasphtm.mid">
<center>
<H1>CIG Sample Web Input Form</H1>
<H3>This Demonstration Calculates Your Approximate Calorie Requirements
</H3>
</center>
```

In these lines of code, we have specified that the data entered into the two text boxes will be posted to the page SampleASP.ASP. The "Submit" button serves the function of executing the POST action. The input text boxes create the variables that are transmitted to the Active Server page.

```
<PRE>
<FORM METHOD=POST ACTION="SampleASP.asp">
<P><B>       Your Weight in Pounds: <INPUT TYPE="text" NAME="WeightVar"
SIZE=10>
<P><B> Minutes of Exercise Per Day: <INPUT TYPE="text" NAME="ExerciseVar"
SIZE=10>
</PRE>
<INPUT TYPE="Submit" VALUE="Get The Answer">
<INPUT TYPE="Reset" VALUE="Prepare For New Data">
</BODY>
</HTML>
<hr>
```

Here is an element that I find missing all to often on Web sites that I visit—a link to the home page. Please give your guests a method of navigating in your site:

```
<a href="default.htm">Return to Home Page</a>
<hr>
</BODY>
</HTML>
```

And now a look at the Active Server page SampleASP.ASP that does the work. The first line declares the script language:

```
<%@ LANGUAGE="VBSCRIPT" %>
```

These lines are the same as before:

```
<HTML>
<HEAD>
<META NAME="GENERATOR" Content="Microsoft Visual Interdev 1.0">
<META HTTP-EQUIV="Content-Type" content="text/html; charset=iso-8859-1">
<TITLE>Active Server Answer Page</TITLE>
</HEAD>
<BODY BGCOLOR = "#FFFFFF">
<bgsound src="images/sampleaspasp.mid">
<center>
<H1>Active Server Answer</H1>
<h3>Your Approximate Calorie Requirements</h3>
</center>
<hr>
```

Here is the VBScript that performs the work of answering the question. It receives the variables from the HTML form, and then calculates and prints the answer:

```
<%Weight = Request.Form("WeightVar")
  Exercise =  Request.Form("ExerciseVar")
  Calories = (Weight * 15) + (Exercise * 5)%>
Your Estimated Daily Calorie Requirement Is
<% Response.Write(CStr(Calories))%>
 Eat Hearty!
```

And a way back to the home page:

```
<hr>
<a href="default.htm">Return to Home Page</a>
<hr>
</BODY>
</HTML>
```

As you are beginning to see, all of this is quite simple. Granted that the Web pages here are simple by design, but the creation of a Web site is a fairly straightforward matter when you are using Visual InterDev. It keeps track of details and provides facilities that shorten the time to perform tasks.

Adding a Client Script Page

The application shown in the following figure is basically useless because you can and do most of these conversions in your head. With a little imagination, you could provide a service to your visitor that would keep them returning to your site. As an example, if you are in real estate, you could have a page such as this that would calculate closing costs or mortgage payments. Yes, I could have created such a page but I promised to keep it simple. (The name of this file is ClntScrptPg.HTM.)

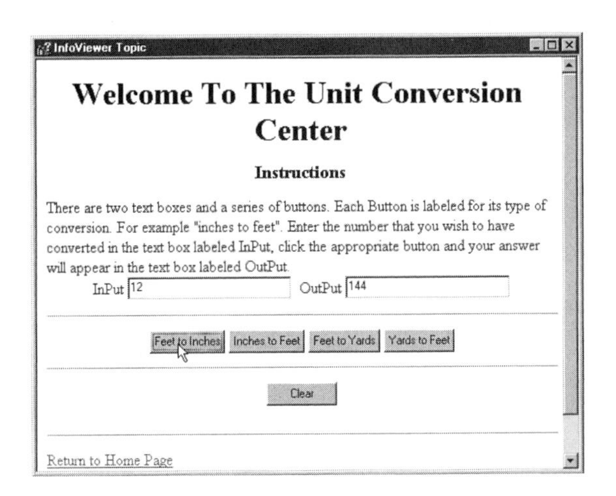

If you leave the left text box blank and press any button, you will generate a type mismatch error. Error handling has been omitted from these pages. Error handling is a necessary element.

Now for a look at the HTML code to see if there are any new developments. The first part we have seen before.

```
<HTML>
<HEAD>
<META NAME="GENERATOR" Content="Microsoft Developer Studio">
<META HTTP-EQUIV="Content-Type" content="text/html; charset=iso-8859-1">
<TITLE>Sample Client Script Page</TITLE>
</HEAD>
<BODY BGColor=#FFFFFF>
<BGSound src="images/clntscrptpg.mid" loop=1>
<Center><h1>Welcome To The Unit Conversion Center</h1>
<h3>Instructions</h3></center>
There are two text boxes and a series of buttons.
Each Button is labeled for its type of conversion.
For example "inches to feet".
Enter the number that you wish to have converted in the
text box labeled InPut, click the appropriate button
and your answer will appear in the text box labeled OutPut.
<br><center>InPut <input type="text" Name="TextBox1" size="30">
OutPut <input type="text" Name="TextBox2" size="30"><BR></center>
<hr>
```

```
<center>
<Input type=Button Name="cmdButton1" Value="Feet to Inches">
<Input type=Button Name="cmdButton2" Value="Inches to Feet">
<Input type=Button Name="cmdButton3" Value="Feet to Yards">
<Input type=Button Name="cmdButton4" Value="Yards to Feet">
<br><hr><input type="Reset" Value="Clear"></center>
<br><hr>
<a href="default.htm">Return to Home Page</a><br><hr>
```

This is a VBScript block that has created four event handling subroutines. There is one event handler each for the four command buttons. The event being handled is the `OnClick` event:

```
<Script Language="VBScript">
     Sub cmdButton1_OnClick
          TextBox2.Value = TextBox1.Value * 12
     End sub
     Sub cmdButton2_OnClick
          TextBox2.Value = TextBox1.Value / 12
     End sub
     Sub cmdButton3_OnClick
          TextBox2.Value = TextBox1.Value / 3
     End sub
     Sub cmdButton4_OnClick
          TextBox2.Value = TextBox1.Value * 3
     End sub
</script>
</BODY>
</HTML>
```

The Script Wizard

One issue that arises when you are writing VBScript or JScript is how do you know what events and properties are available to you?

Visual InterDev has an answer—the Script Wizard. The Script Wizard is opened by right-clicking an HTML form in the text edit window. Choose **Script Wizard** from the menu. (You should create the buttons and test boxes that you are going to use before you open the Script Wizard.) As you see in the next figure, the events are exposed on the left and the properties on the right. It is a simple matter to generate syntactically correct script with this tool.

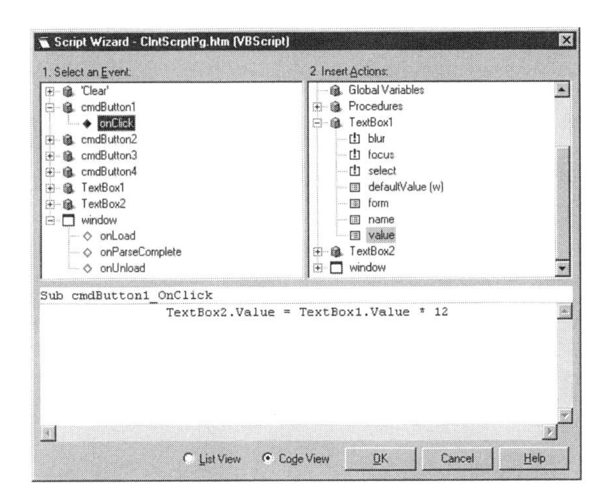

Events that have script generated will show as a filled diamond.

Working with Visual InterDev, you get the feeling that when voice recognition becomes a viable reality, all you will need to do is provide a vague description and Visual InterDev will do the rest. But not quite yet.

Connect to a Database

An Active Server page was used to connect to a database using ActiveX Data Object. The main reason is that ADO only works with Active Server pages. This is because ADO uses ODBC and must be on the same system as ODBC (which is the server). Your next question is "How do I get parameters from the user to limit the Query?" Thanks for asking. We saw in a prior section how we can get input from a user through an HTML form. The same technique will work here. In the figure below, you see the listing from my fictional phone book database.

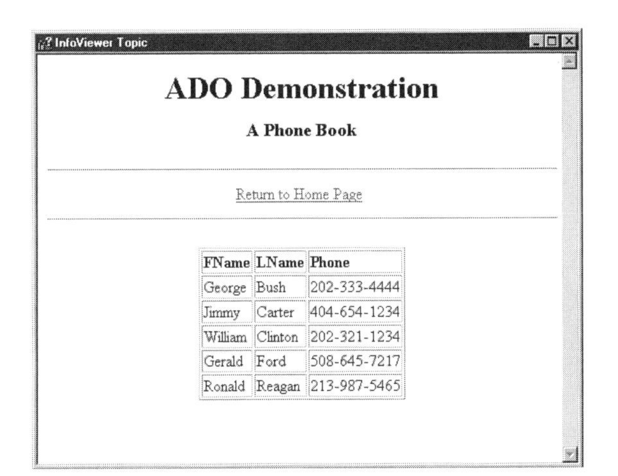

FName	LName	Phone
George	Bush	202-333-4444
Jimmy	Carter	404-654-1234
William	Clinton	202-321-1234
Gerald	Ford	508-645-7217
Ronald	Reagan	213-987-5465

The table in this Web page is formatted using some marvelous code that was borrowed from an example included with the Active Server RoadMap examples.

The first section we have seen before:

```
<%@ LANGUAGE="VBSCRIPT" %>
<HTML>
<HEAD>
<META NAME="GENERATOR" Content="Microsoft Visual Interdev 1.0">
<META HTTP-EQUIV="Content-Type" content="text/html; charset=iso-8859-1">
<TITLE>Active Data Object</TITLE>
</HEAD>
<BODY bgcolor=#FFFFFF>
<bgsound src="images/db_ado_page.mid">
<center>
<H1>ADO Demonstration</h1>
<h3>A Phone Book</h3>
<hr>
<a href="default.htm">Return to Home Page</a>
<hr>
```

Here is the VBScript code that creates the ADO object and establishes the connection. The connection uses an ODBC Data Source Name (DSN) that has already been established. When the SQL statement is executed, the results set is available for the next section of code:

```
<%
Set Conn = Server.CreateObject("ADODB.Connection")
Conn.Open "CIG_PhoneBook_ADO"
Set RS = Conn.Execute("SELECT FName, LName, Phone FROM PhoneBook Order by
LName")
%>
```

This section processes the results set and creates the HTML table for display on the browser. I have used this table code a number of times. Make a copy and keep it. It works very well:

```
<P>
<TABLE BORDER=1>
<TR>
<% For i = 0 to RS.Fields.Count - 1 %>
     <TD><B><% = RS(i).Name %></B></TD>
<% Next %>
</TR>
<% Do While Not RS.EOF %>
```

```
        <TR>
        <% For i = 0 to RS.Fields.Count - 1 %>
        <TD VALIGN=TOP><% = RS(i) %></TD>
        <% Next %>
        </TR>
<%
        RS.MoveNext
Loop
```

This code closes the connection and cleans up. Don't omit it or you will learn about MEMORY LEAK. (You will be required to write it on the blackboard 100 times.)

```
RS.Close
Conn.Close
%>
</TABLE>
</BODY>
</HTML>
```

Amazingly few lines of code are needed to accomplish such a complex task. It is getting easier. Thank you Microsoft.

HTML Layout

The use of the HTML Layout Control requires the use of an .ALX page. The HTML Layout Control is an ActiveX Control that provides a precise arrangement of elements, as you can see in the figure below.

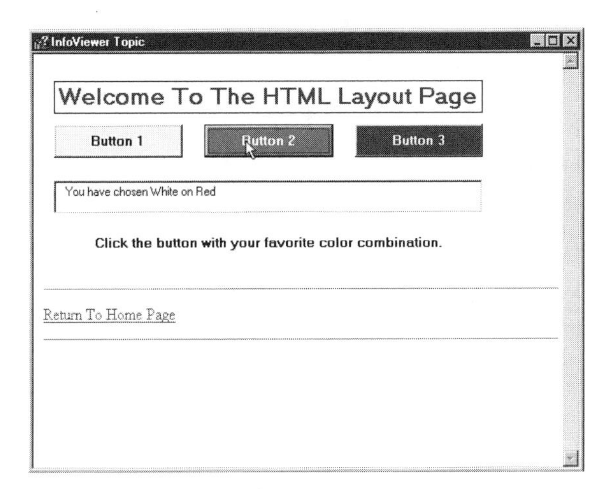

The Layout Control can be combined with standard HTML elements on a page. The link to the home page is not part of the Layout Control.

When you look at the HTML on the file LayoutPage.HTM, you see that the primary task performed is the insertion of the HTML Layout object. Note that the object refers to another file—LayoutPage.ALX. This file contains the parameters, button, and text box objects that are included in the Layout Control:

```
<HTML>
<HEAD>
<META NAME="GENERATOR" Content="Microsoft Developer Studio">
<META HTTP-EQUIV="Content-Type" content="text/html; charset=iso-8859-1">
<TITLE>Document Title</TITLE>
</HEAD>
<BODY BGCOLOR=#FFFFFF>
<BGSound src="images/layout.mid" loop=1>

<OBJECT CLASSID="CLSID:812AE312-8B8E-11CF-93C8-00AA00C08FDF"
ID="Html_Layout1" STYLE="LEFT:0;TOP:0">
<PARAM NAME="ALXPATH" REF VALUE="LayoutPage.alx">
</OBJECT>

<hr>
<a href="default.htm">Return To Home Page</a>
<hr>
</BODY>
</HTML>
```

Scripting for an HTML Layout Control

If you are going to perform any scripting that involves the HTML Layout Control, the scripting must be on the .ALX file. OK, you ask "How do I open the .ALX file in a text editor?" Thanks for asking. Right-click the file name in the Workspace Window, choose **Open With** from the menu. Select **Source Editor** in the Open With dialog box and click the **Open** button. You will see the text version of the HTML Layout Control .ALX file.

The VBScript to work with the buttons and text boxes in the Layout Control can be seen in the .ALX file on the CD. As you scan through the code, you will recognize the elements of the Layout Control.

ActiveX

ActiveX and Browser Security

Before you begin to work with ActiveX Controls, check the Security setting of your Microsoft Internet Explorer 3.0. You can do this by right-clicking the Internet Explorer icon on your desktop and choosing properties. Click the **Security** tab and click the **Safety Level** button. Select the **Medium** option button. If you are set on High, your browser refuses to allow ActiveX Controls from an unknown source.

ActiveX Controls can be integrated with standard HTML. The Calendar Control in the center of the form has HTML elements above and below. Also, the elements at the bottom of the page interact through VBScript with the ActiveX Calendar Control.

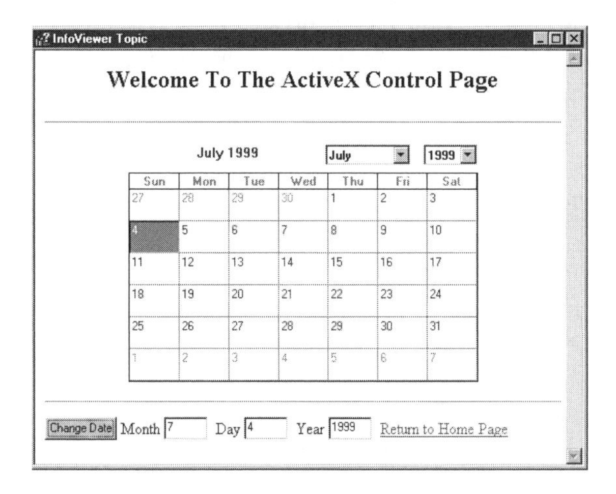

Prepare for fireworks. The date is July 4, 1999.

When you examine the HTML for this page, you find that the use of the ActiveX Control makes the page more compact than it would be if all of the functionality were in VBScript. Notice the Sub window_onload. This sets the calendar to the current system date. The other Sub cmdButton1_OnClick allows setting the date from outside the control.

```
<HTML>
<HEAD>
<META NAME="GENERATOR" Content="Microsoft Developer Studio">
<META HTTP-EQUIV="Content-Type" content="text/html; charset=iso-8859-1">
```

```
<TITLE>Sample Site ActiveX Control Page</TITLE>
</HEAD>
<BODY bgcolor=#FFFFFF>
<BGSound src="images/activex.mid">
<Center><h2>Welcome To The ActiveX Control Page</h2></center>
<hr>
<Center>
    <OBJECT ID="Calendar1" WIDTH=372 HEIGHT=250
     CLASSID="CLSID:8E27C92B-1264-101C-8A2F-040224009C02">
        <PARAM NAME="_Version" VALUE="458752">
        <PARAM NAME="_ExtentX" VALUE="9843">
        <PARAM NAME="_ExtentY" VALUE="7382">
        <PARAM NAME="_StockProps" VALUE="1">
        <PARAM NAME="BackColor" VALUE="16777215">
        <PARAM NAME="Year" VALUE="1997">
        <PARAM NAME="Month" VALUE="2">
        <PARAM NAME="Day" VALUE="8">
        <PARAM NAME="DayFontColor" VALUE="255">
    </OBJECT>
</center>
<hr>
    <INPUT TYPE=Button VALUE="Change Date" NAME="cmdButton1">
Month
    <INPUT TYPE=Text SIZE=5 NAME="TextBox1">
Day
    <INPUT TYPE=Text SIZE=5 NAME="TextBox2">
Year
    <INPUT TYPE=Text SIZE=5 NAME="TextBox3">
<a href="default.htm">Return to Home Page</a>
    <SCRIPT LANGUAGE="VBScript">
Sub window_OnLoad
     Calendar1.Month = Month(date())
     Calendar1.Year = Year(date())
     Calendar1.Day = Day(Date())
End sub
Sub cmdButton1_OnClick
     Calendar1.Month = TextBox1.Value
     Calendar1.Day = TextBox2.Value
     Calendar1.Year= TextBox3.Value
End sub
```

```
    </SCRIPT>
</BODY>
</HTML>
```

ActiveX Controls are a compact method of extending the functionality of a Web page. With Visual Basic 5.0 Control Creation Edition, you can create your own ActiveX Controls.

The Relaxation Room

Okay, it does nothing but play some bad music, but the picture is pretty. It is a great place to relax. Sort of like listening to MUZAK for entertainment. The file RelaxPage.HTM will show some issues of color resolution if your Windows display is set to 16 colors.

Search

The Search.HTM page is the standard search page created by Visual InterDev with the following HTML added. Use the examples that are provided by Visual InterDev. Make changes to suit your needs. It saves a lot of time, effort, and frustration:

```
<body BGColor=#ffffff>
<bgsound src="images/search.mid" Loop=1>
<h1><a name="top">Sample Site Text Search</a></h1>
<hr>
<a href="default.htm">Return to Home Page</a>
<br>
<hr>
```

Run the Search page to see what it does. It is really neat.

Check the Links

After you have your created Web site, you will want to check the links to be sure that they all perform. There are two ways to do this, the hard way and the easy way. The hard way is to trace all of the links by exercising them. The easy way is to click the **Default.HTM** file name in the Workspace window and choose **View Links**. You will see a Link View edit window as shown in the following figure. The links with an arrow head on both ends indicate a two-way link. The one broken link in the figure was because the modem connection to the Internet was not open.

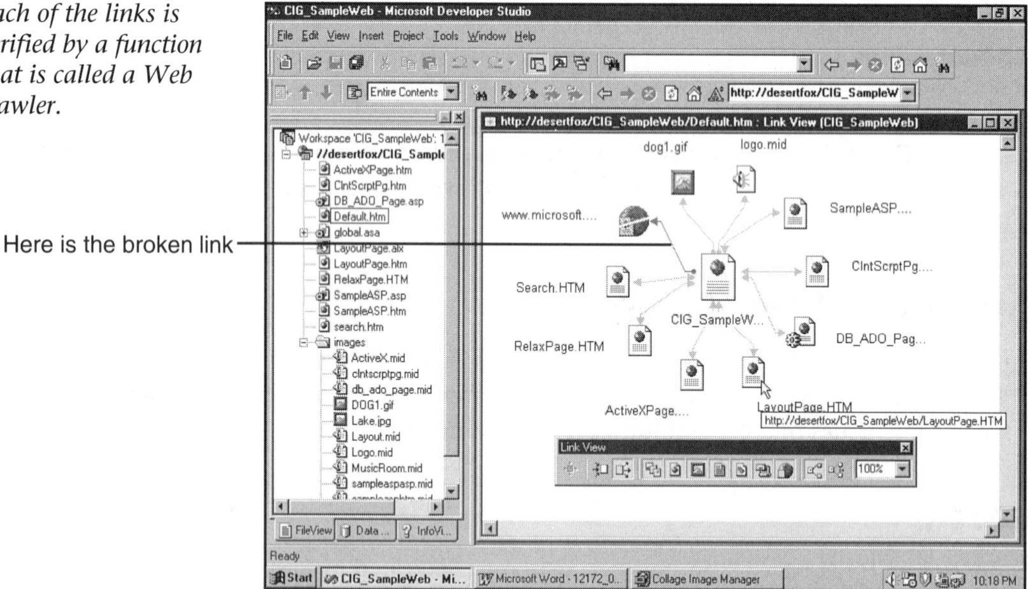

Each of the links is verified by a function that is called a Web crawler.

The Least You Need to Know

This chapter has been a fast trip through a vast array of topics. The intent was not to explain everything in depth. (That is done much better in other chapters.) This chapter is intended to pull a lot of issues together so that you can see them in relation to each other and see how they can be integrated. You should understand the process of creating a Web site a little better now. While it is a complex process, it is also a lot of little issues that need to be managed. This is where Visual InterDev is of such assistance. All of the tools are there and all of the details are remembered by Visual InterDev. Just remember:

- The best method of learning Visual InterDev is to *develop* using Visual InterDev. You can read about it until your eyes are black sockets and you will still not have the understanding that comes from use.
- Visual InterDev is a tool, not a goal. Your goal is the development of Web sites using Visual InterDev. Visual InterDev is only the bus that takes you to your goal, not the destination itself. (Actually, Visual InterDev is more of a jet fighter than a bus.)
- Visual InterDev is a very broad development environment. There are many ways to accomplish a given goal with Visual InterDev. Don't think that the methods that you see in this book are the only ways to approach a task or are the best for you. The goal of this book is to get you started. There is more gold to be found.

Part 7
Final Issues

Microsoft Visual InterDev operates on three Microsoft operating system platforms: Windows 95, Windows NT 4 Server, and Windows NT 4 Workstation. It interfaces with three Microsoft Web servers: Personal Web Server for Windows 95, Peer Web Server for Windows NT 4 Workstation, and Internet Information Server for Windows NT 4 Server. In addition, Microsoft FrontPage has a Web server that will operate on all of the Microsoft 32-bit operating systems.

The differences and similarities of these various combinations deserve some examination. That is what will be covered in this final section.

NEW AIR TRAVEL SAFETY LECTURE

Chapter 25

My Web Server's Better Than Your Web Server

In This Chapter

- ➤ Explore the capabilities of Microsoft Internet Information Server and Microsoft Personal Web Server
- ➤ Examine the differences between the two Web servers

Visual InterDev will function equally well with three different Web Servers. Which of the three Web servers you are using will have little impact upon your Web development as performed with Visual InterDev. The differences in the Web servers is in their capabilities as delivery vehicles, not in the Webs that they will serve. The main difference is the fact that ISAPI support is not available in the Personal Web Server as it is in MS IIS. Since ISAPI is not a tool that is used with Visual InterDev, the difference is not significant for our purposes. A brief comparison of the capabilities and differences of each of the Web servers will provide a basis for understanding the relationship between Visual InterDev and each Web server.

Check This Out...

When Three are Two There are three Web servers that can be used with Visual InterDev. These are MS IIS, Peer Web Services, and Personal Web Server. The reality is that MS IIS and Peer Web Services are virtually indistinguishable. Therefore, this chapter is a comparison of MS IIS and Personal Web Server.

MS IIS stands at one end of the capability spectrum (most capable) and Personal Web Server stands at the other (least capable); Peer Web Services is basically the same as MS IIS.

Administration

In the figure below, you see the Microsoft Internet Service Manager. The Internet Service Manager is the program that serves as the control center for the MS IIS Web Server. It is opened by choosing **Start**, **Programs**, **Microsoft Internet Server (Common)**, **Internet Service Manager**.

You can manage multiple servers and their services from this interface.

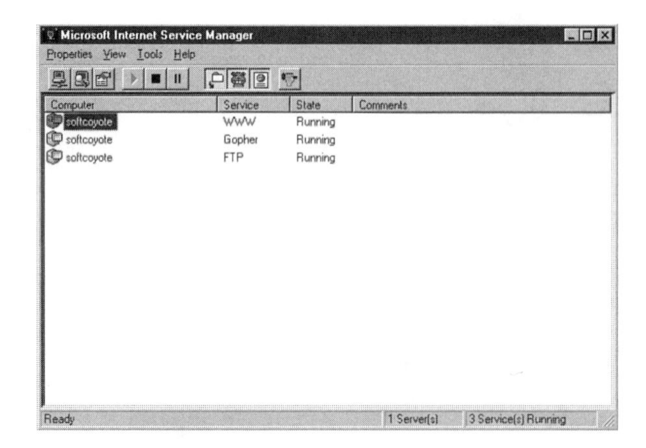

Opening the Server Managers

The MS IIS server may be managed from the local machine through the Microsoft Internet Service Manager or through a Web browser. The Personal Web Server is managed through a Properties dialog box and through a Web Browser.

This installation has three services running, WWW Publishing Service, Gopher Publishing Service, and FTP Publishing Service. There are other services that can be managed through this interface, including the Proxy Server, the Index Server, and the News Server.

In the following figure, you see the Properties dialog box for the Personal Web Server.

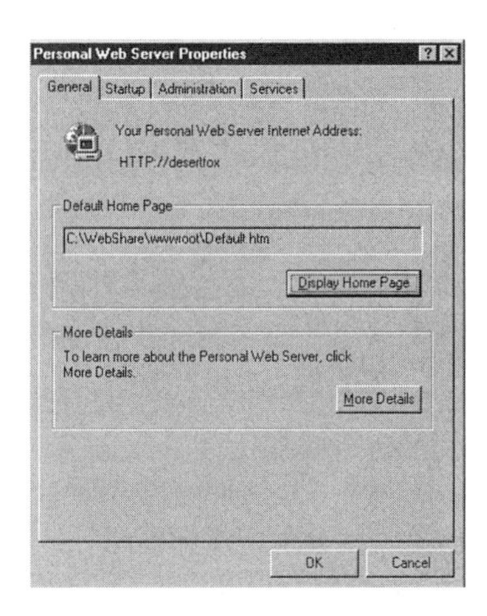

Click the More Details button to see documentation for the Personal Web Server.

Only the local Personal Web Server can be seen and managed.

General Settings for Web Services

The dialog box for the WWW Service Properties for MS IIS as shown in the figure below allows for setting the TCP port and an Anonymous Logon account. The Anonymous Logon account is used by NT Security.

NT Anonymous Account

When MS IIS is installed, an anonymous account with the name of IUSR_ServerName is created and all of the proper permissions are set in the Windows NT Security for the account and the root Web.

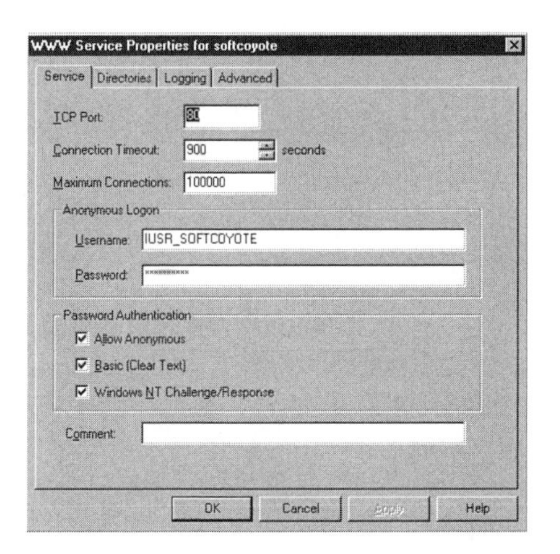

Port 80 is the accepted and most popular port for the World Wide Web.

The Personal Web Server Administrator shown in the next figure has no provision for setting the TCP port. It is always port 80. There is also no provision for an anonymous logon. Windows 95 (on which the Personal Web Server runs) does not have the NT Security services.

The Maximum Connections property will help keep the system from being over-burdened.

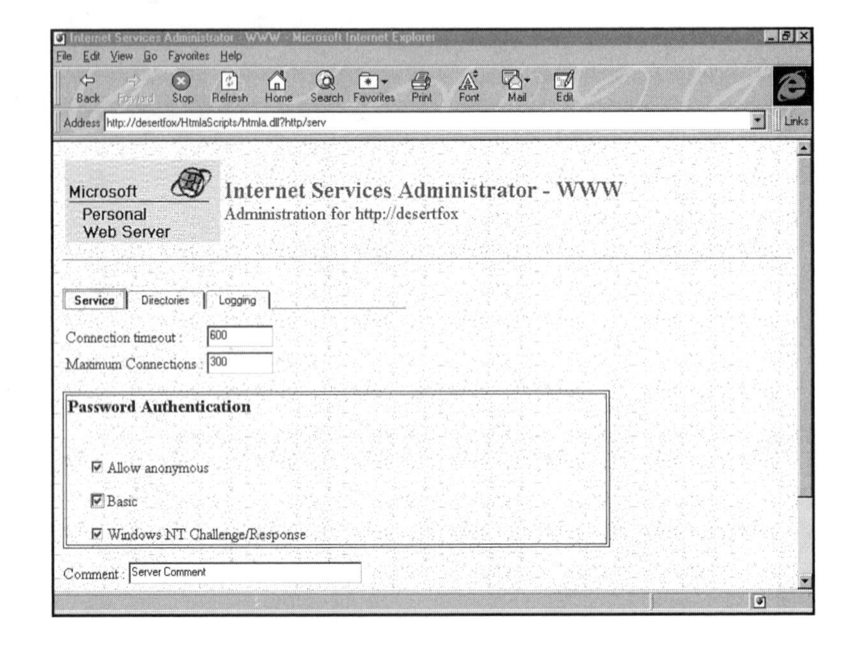

Directories

The settings allowed for directories in general are identical for both MS IIS and the Personal Web Server as is seen in the next two figures. Both servers allow for the use of default documents and directory browsing.

Multiple default document names may be used.

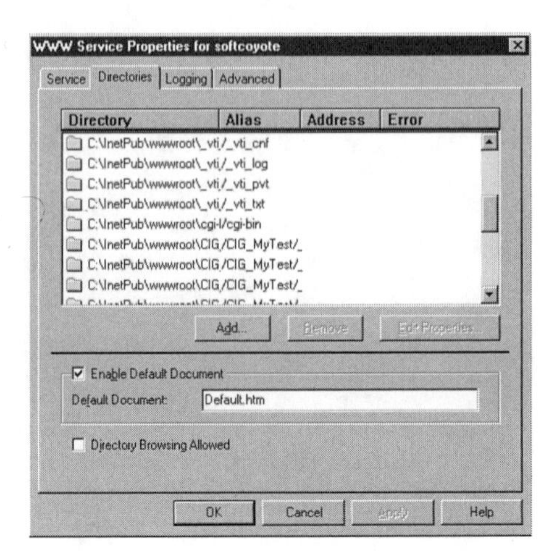

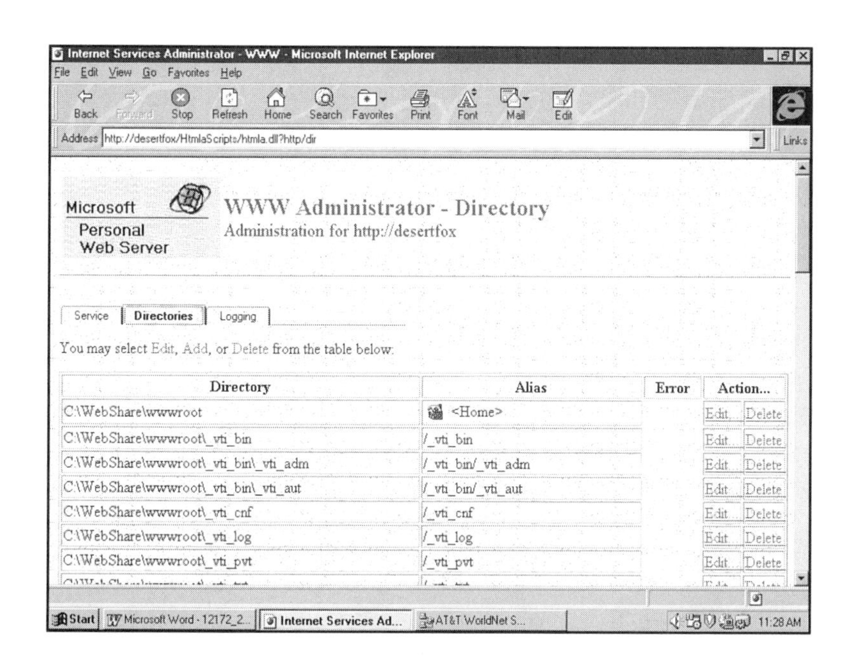

Although the appearance is quite different with the Personal Web Server, the function is identical.

Editing a Directory

When the properties for a specific directory are edited, you see some of the marked differences in capability between MS IIS and Personal Web Server. MS IIS, shown in the figure below, allows for specific account access, a virtual server, Secure Sockets Layer, and Certificates.

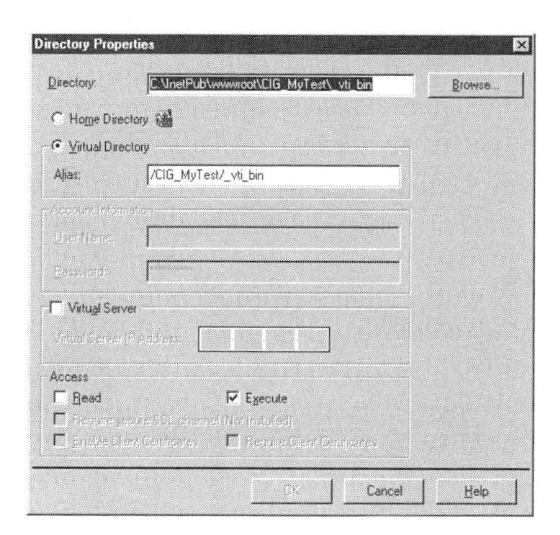

Virtual servers are also called multi-homed servers.

None of these features are available on the Personal Web Server as shown in the following figure.

Both the physical and the virtual directory paths are shown.

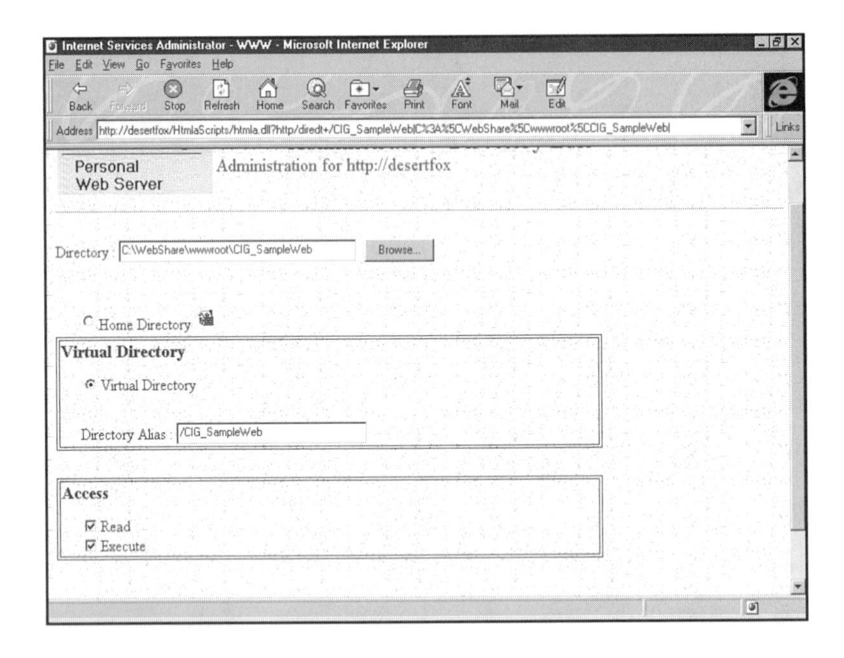

Both of the servers allow the setting of Read and Execute access and the creation of Virtual Directories.

Logging

The same information is logged by both MS IIS and Personal Web Server. The difference as can be seen in the next two figures is that MS IIS allows for logging to an ODBC database while Personal Web Server does not.

MS IIS provides for logging in both Microsoft standard format and NCSA format.

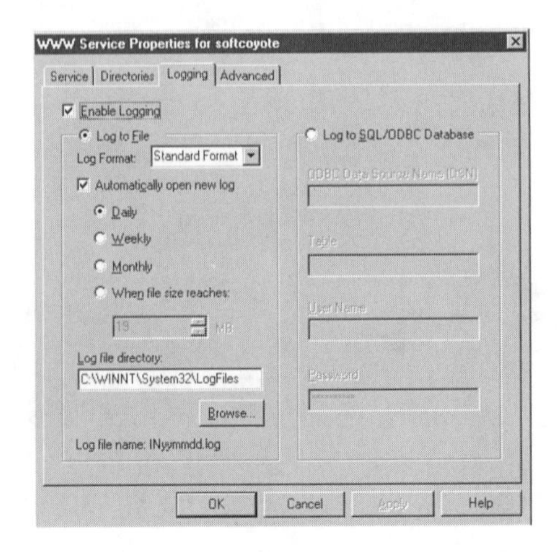

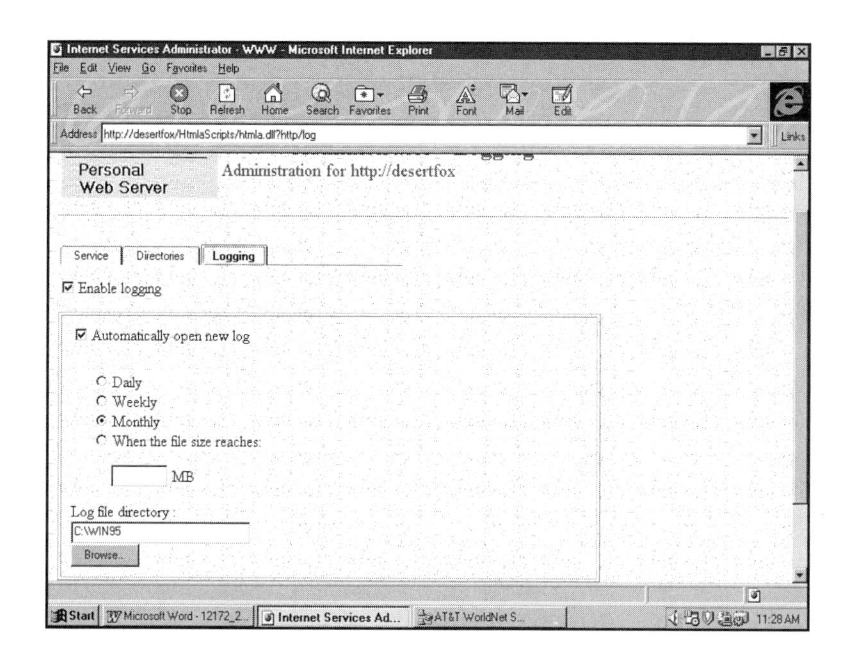

You may keep a daily, weekly, or monthly log.

Limiting Access by IP Address

MS IIS also provides access control by specific IP address as shown in the figure below.

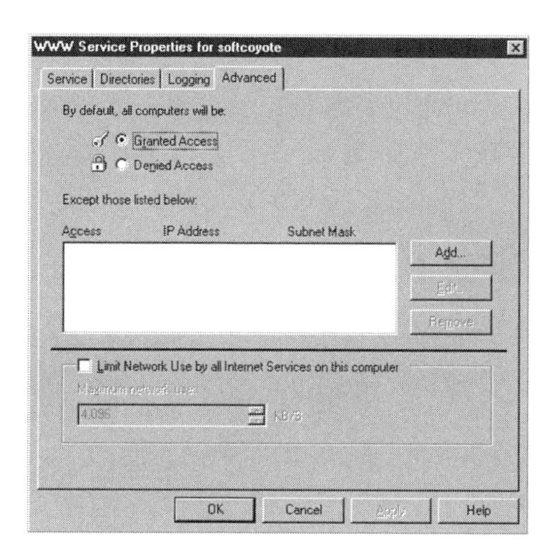

MS IIS also provides a total bandwidth control for access to a server. It limits the entire server.

The Least You Need to Know

This has been a quick tour through the Web Server aspects of MS IIS and MS Personal Web Server. To explore these servers in depth would require a book each. The similarities touch on the important aspects. Because of the similar use of the directory structure, a Web can be developed on the Personal Web Server and moved to MS IIS with no changes. Just remember:

- The differences between these servers are in the areas of security and performance, not Web structure.
- There is no advantage to developing on one server over another. The closest server where you have the easiest access is probably the best choice.

Appendix A

Installation and Setup of Software

There are some issues about the installation and setup of this software that you need to know. You should find the answers to most of your questions here.

In case there is a question to which you don't find an answer here, you should check out the news groups on the news server **msnews.microsoft.com**. (I have found many answers there in the writing of this book.)

Some of the software covered in this book comes in two flavors: Windows NT 4.0 and Windows 95. The difference here is in the Web Server that is installed. There is also a specific order for the installation of some of the various pieces of software.

Be Curious

If you don't find the answer, post a question. And don't feel shy about being a newbie. (This is news speak for someone who is just beginning with a particular piece of software.) If you say that you are a newbie, the chivalry of the regulars comes out. One technique is to announce yourself as a newbie and say that if your question is answered in the FAQ's (Frequently Asked Questions), please point you in the right direction.

The FAQ's are on the Microsoft site **www.microsoft.com**. These are kept up to date and provide wonderful information, most of it coming from the Microsoft Tech Support Group.

Windows 95 Personal Web Server

The setup of the Windows 95 Personal Web Server is very straightforward:

1. Simply place the Visual InterDev CD-ROM in your CD-ROM drive. If it doesn't autorun, go to the Windows 95 Explorer and double-click the **Setup.exe** file in the root directory of the CD-ROM. This will invoke the master setup menu.
2. It will have a list of the software on the CD-ROM. The list is in two parts: the Server Components and the Client Components. At the bottom of the list is an item named Readmes. This contains the installation instructions. When you click this, an HTML document appears in your Web browser.
3. Your Web Browser should be Internet Explorer 3.0 Build 1215 or later. If you don't have this build, it is downloadable from **www.microsoft.com**. It is also on the CD-ROM in the directory Client\IE30. There are two files there. The smaller file is Internet Explorer only and the larger includes the Microsoft Internet Mail and Internet News clients.

Build Number
If you don't know the build number of your Internet Explorer, when it is open, go to Help|About. Part of the information displayed will be the version. The version on mine is Version 3.0 (4.70.1215) The last four numbers are the build.

4. Read the Installation Instructions HTML document in your browser window carefully. It will contain any late-breaking news. After you are sure that you have covered any necessary issues, close the document and browser and click the menu item **Personal Web Server** (Windows 95 Only). At this point, the setup program takes over and does the job with no decisions that have to be made.
5. The installation process creates a directory—Webshare—on your C: drive. After the installation is complete, you will need to restart your computer. When you have restarted, the Personal Web Server will start automatically.
6. You will see a new icon on the lower-right of the Windows 95 taskbar by the System Time display. This is for the Personal Web Server. When you hold your mouse over the icon, a tip box will appear that announces it as the Personal Web Server. When you double-click, the properties dialog box appears as shown in the following figure.

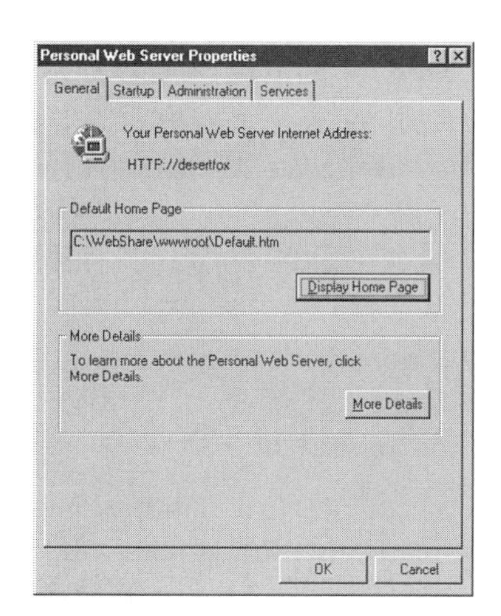

You can control the setting for the Personal Web Server from this dialog box.

The Properties Dialog Box

When you right-click the Personal Web Server icon, a menu appears that provides three choices, one of which is this Properties dialog box. The other two menu choices are Home Page, which opens your Home Page, and Administer. Both of these can be selected from the Properties dialog box.

The properties dialog box has four tabs:

- **The General Tab** Displays information about the Personal Web Server such as the address and the URL. It also provides a button to open your Home Page and get More Details about the Personal Web Server.
- **The Startup tab** Lets you start or stop the Personal Web Server and set two options. The first option is whether to have the Personal Web Server start automatically at the system boot. The second sets whether to display the Personal Web Server on the taskbar. Leave this one set to **yes**. It is difficult to impossible to access these properties otherwise.
- **The Adminstration Tab** Displays one button that opens the Administration pages in your Web site. This is discussed in Chapter 19, "Web Site Security: Guarding the Gate."
- **The Services Tab** Shows that there are two services: HTTP and FTP. It also allows you to set whether the services start automatically or manually.

If you have problems with the browser finding your Web site, the cause may be that the TCP/IP stack is not running. An easy way to get your TCP/IP stack running is to let Internet Explorer make the usual connection to your Internet Service Provider.

Windows NT 4.0 Internet Information Server

There are actually two Web servers for Windows NT 4.0. The Internet Information Server runs on the NT Server and the Peer Web Server runs on NT Workstation. These instructions and illustrations are based on IIS 2.0 for NT Server 4.0. (The Peer Web Server is so similar that you will have no problems using these instructions. It is essentially the same product as MS IIS 2.0).

When you installed Windows NT 4.0 Server, you are offered the opportunity to install the Microsoft Internet Information Server 2.0. If you did so, you can skip this section. If you now need to install IIS 2.0, read on:

1. Place the Windows NT 4.0 Installation CD-ROM in your CD-ROM Drive. If the CD-ROM starts automatically through autorun, click **Browse This CD**. If it doesn't start automatically, open the CD-ROM in the Windows NT Explorer.
2. Open the correct directory for the platform of your system, such as I386 for an Intel based system. Open the Inetsrv directory and start the program Inetstp.exe.
3. After the Welcome dialog box, you are presented with a dialog box where you choose the parts of IIS that you want installed. The minimum that you should choose are the Internet Service Manager, World Wide Web Service, and the ODBC Drivers and Administration.
4. You are then presented with directory assignment dialog boxes. After these have been completed, IIS is installed. You are asked about the ODBC Drivers to be installed. Don't be concerned about overwriting a newer driver with an older driver. The version checking works.
5. To find the administration program for MS IIS, choose **Start, Programs, Microsoft Internet Server (Common), Internet Service Manager**. The Web site is in a directory Inetpub on your C: drive.

Reviewing the administration of MS IIS is beyond the scope of this book. If you want more information on MS IIS, there is an excellent book by Que, *Special Edition Using Microsoft Internet Information Server 2*. (Shameless commercial warning: I wrote some of the chapters of the book).

Microsoft Active Server

The Active Server Pages is the second choice on the Server Components menu. Click this choice and the setup will start.

1. If the Personal Web Server is running (or for Windows NT, IIS is running), the setup program will tell you that it must be stopped for the installation to continue and ask if you want to stop it. Answer **Yes**.

2. You are then presented with a dialog box that asks which options you want installed. You will want all of the options that include the Active Server Core Pages, ODBC 3.0 + Access Driver, On-line Documentation, and the Java VM (Virtual Machine).
3. Next you are presented with a directory assignment choice. Accept the default. The setup program then completes the installation.
4. You are presented with a dialog box telling you where everything was installed. Click **OK**.
5. Next, you are informed that a choice was added to your Microsoft Personal Web Server Group called Active Server Pages Roadmap. This Roadmap is a source of valuable information. Visit it soon to see what is available.
6. You are next offered the choice of restarting your system. You should restart as soon as you have closed any other programs and saved any work.

Check This Out...

The Road Map
The Roadmap provides information on VBScript, JScript and excellent examples of the use of Active Server pages. Good documentation is also available here on all aspects of Active Server.

FrontPage Server Extensions

FrontPage Server Extensions are the third choice on the Server Components Menu. To install the Server Extensions:

1. Click this choice and the setup program starts. It first informs you that it must stop the Personal Web Server or MS IIS before it can install. Choose **Yes**.
2. You are next presented with a directory assignment dialog box. Accept the default.
3. Next you are asked if you want Typical or Custom install. **Typical** is a safe choice.
4. You are then asked on which server you want the FrontPage Extensions installed. Accept the Personal Web Server or IIS.
5. Finally you are informed the World Wide Web Service must be restarted to complete the installation. Choose **Yes**.
6. You may now be presented with information regarding User and Share level access control. Click **OK** and accept the final announcement that the task is completed and you are finished.

Microsoft Visual InterDev Client

Since the Installation Instructions cover the Client Components, you are ready to install the first item in the list, Microsoft Visual InterDev Client. Click the second item in the list and the installation program starts. The only choice that you are offered in the installation that you need think about is whether you want a Typical, Custom, or Compact installation. For working with this book, I would recommend the Custom option and that you select all components. It does not require a great deal of additional disk space. It adds the Books-online, which you will find invaluable. It also adds the sample projects, which you will use with this book. After this, the installation program takes over and you are finished.

Microsoft Image Composer

The Image Composer is a separate application from Visual InterDev. To start the installation, click the Menu item **Microsoft Image Composer** on the **Master setup Client Components** menu. As you go through the installation, you are offered the choice of Typical, Custom, or Compact. I recommend that you select the Custom so that you can see the components. If you select Typical, the items that won't be installed on your hard disk are the Font Samples 2 Megs, Photo Art 237 Megs, and Web Art Samples 24 Megs. If you choose not to install these components, they are still available to you on the CD-ROM. The path is **\Client\ImgComp\MMFiles**. You will find the Photo directory and the Photos, Font, and Web directories there.

You have completed the setup of Microsoft Image Composer. Chapter 5, "Toolbars/Menus/Help and Documentation: What Does this Do?," discusses how you can add Image composer to your Visual InterDev Tools menu.

Microsoft Media Manager

Media Manager is also a separate application from Visual InterDev. Choose the **Master Setup** menu, then choose the **Client Components** list and then choose the **Microsoft Media Manager**. The install program starts the installation. Your first choice is to do a complete installation. Next you are prompted to choose the Annotation database. Unless you are working as part of a Workgroup, choose the Create Default Annotations Database.

Microsoft Media Manager then finishes the installation. When the installation is finished, you are asked to Restart Windows. After you have closed and saved work from any other programs, restart Windows.

Before You Compose One thought: unless you have a sound card on your system, you won't be able to use the Music Generator.

Microsoft Music Generator

Choose the Master Setup Menu, the Client Component, and finally Microsoft Music Generator. You are offered the choice of Typical or Custom Installation. Choosing the Custom installation will show you the disk space required. You will want to install all components. After you have completed this choice, the setup program does the rest.

CONGRATULATIONS. You have installed everything necessary —and even desirable—for working with Microsoft Visual InterDev.

Appendix B

What's on the CD?

Now that you've read the book, there's probably still information you need—but where do you go from here? May I suggest you take a look at the wealth of information and tools the good folks at Que have provided on the companion CD (in case you were wondering what that thing in the back of the book was).

What will you find there? Lots of stuff! So pour yourself a cup of coffee (or your favorite beverage) and prepare for a fun trip into the *Complete Idiot's Guide to Microsoft Visual InterDev*—the CD! Starring, in no particular order:

- **ActiveX Tools** Visual Basic Control Creation Edition 5.0, several cool ActiveX Controls you can insert into your Web pages, Microsoft ActiveX Control Pad, Microsoft ActiveX Redistributable Components, Microsoft ActiveX Software Developer's Kit with Reference Files, Microhelp OLEtools 5, MVB References.
- **Microsoft Internet Assistants** Including those for Excel, PowerPoint, Word, and Access.
- **Microsoft Viewers** Including those for Excel, PowerPoint, and Word.
- **Microsoft Internet Explorer 3.01** With NetMeeting and HTML Layout Control.
- **Multimedia/Virtual Reality Tools and Players** Including mBED, Sizzler, Surround Video, VDOLive Video Player, VRscout, WIRL, and Viscape.
- **Lots of other tools to help you build the most exciting Web pages** Including Crescent Internet ToolPak, Crystal Reports Viewer, SocketWrench, WinZIP, InterAct, IntraApp, Adobe Acrobat Viewer, NetList,TapiDial, RasDial, and Sax Canvas Control.

But wait! There's more fun to be had! In addition to all of the above, don't miss these exciting additions:

- HTML versions of two best-selling Que books that will get you up to speed on Web page design: *Special Edition, Using HTML 3.2* and *The Complete Idiot's Guide to JavaScript, Second Edition.*
- Lots of graphic and sound files you can use to enhance your Web pages (free of charge, of course!).

Sample files must be installed in the Web! In order to get samples to work, you must install them in the Web. The easiest method of doing this is through Visual InterDev. Again, read the "Readme" text file for more information specific to each chapter's examples.

But wait! There's still more! You will also find code files for the samples used in this book (including a complete sample site created with Visual InterDev—see Chapter 24 for more information on this). The files are organized by chapter in the directory Que\Examples. There is a Readme text file on the "Examples" page that will provide instructions for the process of each chapter.

These example files are here for your use. I've made them intentionally simple so that you will be able to see what is being done, *sans* clutter. One use of the examples is to compare your results to mine (to see, as you read through the book and work out the examples on your own system, if you get the same results as mine). Remember, if you get the result that you want, even if your solution is different, you are not wrong. There is only code that works and code that doesn't. There is no right and wrong, so experiment and have fun. Good luck!

Legal Stuff

By opening this package, you are agreeing to be bound by the following agreement:

This software product is copyrighted, and all rights are reserved by the publisher and author. You are licensed to use this software on a single computer. You may copy and/or modify the software as needed to facilitate your use of it on a single computer. Making copies of the software for any other purpose is a violation of the United States copyright laws.

This software is sold *as is* without warranty of any kind, either express or implied, including but not limited to the implied warranties of merchantability and fitness for a particular purpose. Neither the publisher nor its dealers or distributors assumes any liability for any alleged or actual damages arising from the use of this program. (Some states do not allow for the exclusion of implied warranties, so the exclusion may not apply to you.)

Appendix C

Speak Like a Geek: The Complete Archive

.ALX The file extension for the text file that is used by the HTML Layout Control. This file contains the parameters for the HTML Control. It also contains any script for the HTML Layout Control.

.ASA The file extension for the Global.ASA file. This is an Active Server file. There is only one Global.ASA file for each Web, and it is always located in the Web root directory. It is used to process event handlers for events that occur at the start of the application or a session.

.ASP The file extension for an Active Server Page. This extension signals the Active Server to read the page and generate HTML to be sent to the browser rather than just sending the page.

.GIF The file extension for a graphic image format that is used extensively on Web pages on the Internet. It has the advantages of being supported by most Web browsers and being a compressed format. See *.JPG*.

.HTM The file extension for an HTML page.

.JPG The file extension for a graphic image format that is used extensively on Web pages on the Internet. It has the advantages of being supported by most Web browsers and being a compressed format. See *.GIF*.

.MIC The file extension for the native image format of graphic files in Microsoft Image Composer. The image is composed of a background and one or more sprites. See *Sprites*.

.MID The file extension for a MIDI sound file. It is supported by most Web browsers. Microsoft Music Producer can save files in this format for use in Web pages.

.MMP The file extension for the native sound file format for Microsoft Music Producer.

Active Server Pages HTML pages that are read by the Active Server and an HTML file is generated to be sent to the client browser based on the script language instructions in the Active Server Page.

ActiveX The latest descendent of the COM, OLE family. ActiveX Objects are controls that can be transferred to the browser for processing. They are similar to Java applets.

ALX File The text file that is used by the HTML Layout Control. It is based on a new draft HTML syntax published by the W3C. This file contains the parameters for the HTML Control. It also contains any script for the HTML Layout Control.

ASA The file extension for the Global.ASA file. This is an Active Server file. There is only one Global.ASA file for each Web. It is always located in the Web root directory. It is used to process event handlers for events that occur at the start of the application or a session. There are four events that can be handled by the Global.ASA file. These are the Session_OnStart, Session_OnEnd, Application_OnStart, and Application_OnEnd.

ASCII Text American Standard Code for Information Interchange. This is a 7-bit code for the characters, numbers, and special characters displayed and printed by a computer. One of the primary characteristics of ASCII text is that it can be easily read by humans, unlike a binary file.

ASP The file extension for an Active Server Page. This extension signals the Active Server to read the page and generate HTML to be sent to the browser rather than just sending the page. If script is included in the HTML page, it is processed by the server, not the browser.

Bottleneck Function When an application processes data through multiple functions, the bottleneck function will be that function that is the slowest and will delay the processing. If an application is distributed over multiple computer systems, the bottleneck will be the slowest process on which the other systems must wait.

Broken Link The World Wide Web runs on links. These links can be from one document to another, from one location in a document to another, from one Web site to another. When a Web browser is given a link to process, it generates a request for the target file. A broken link is when the file is not at the URL address specified by the link.

Bytecode The Java language is called a compiled and interpreted language. This is because Java source code is compiled into bytecode which is then iterpreted by the Java Virtual Machine. The Java Virtual Machine will be specific for the computer system platform on which it is running.

C++ A full-featured programming language. C++ is object-oriented and provides classes upon which much of the current development depends. C++ is based on C, an earlier language created for programming on the UNIX operating system computers.

CGI Common Gateway Interface. It is a standard for programs to interface to Web servers. Much of the CGI programming is performed using C script and PERL script.

ClassID Each ActiveX object is assigned a unique identifier as an OLE class object.

CLSID The name of the ClassID property as used in the registry.

Code Context This refers to the manner in which a variable in a loosely-typed language decides what the type of variable really is. In VBScript for example, all variables are of the variant type. Each variant variable will belong to a data subtype such as integer or string. The way the variable is used in the Code Context will determine its subtype. If a variable is initialized with the value of 100, its subtype will be integer. If it is initialized with a value of "Smith" it will be the string subtype.

COM Component Object Model. This is the paradigm upon which OLE and ActiveX are based. Objects in two different processes are able to communicate in OLE.

Compiled A computer language is compiled when the source code, usually a text file, is processed by a program called a compiler. The compiler reads the text instructions written by the programmer and produces machine language instructions that the computer can understand.

Conditional Logic Program logic that tests a condition and varies the path of the program based on the results of the test. The classic conditional statement is the if...then statement.

Database Engine This is the program or programs that perform the database processing. It adds, updates, deletes, and retrieves database records.

DataView Tab This tab of the Workspace window displays the database information in Visual InterDev when a project contains a database connection.

DCOM Distributed Component Object Model. DCOM is the concept of COM with the added feature that the objects communicating can be on different computer systems and communicate transparently.

DEC VT-100 When a workstation acts as a terminal on the Internet in a Telnet session (or any other time for that matter) it will emulate a terminal. The DEC VT-100 emulation is a very common terminal emulation that is used.

Default Page When a request is made for a file from a Web server, if no file is specified, the server transmits the default page if it exists. All servers have a name or names that may be used for default pages. In MS IIS, Peer Web Services, and Personal Web Server, the default page name can be set by the user to any name. The common default page names are Default.HTM, Default.ASP, Index.HTM, and Index.HTML.

Delete Query A SQL statement that will delete one or more records from a relational database table. The records deleted are determined by the WHERE clause in the SQL statement.

Deployment Web In Microsoft Visual SourceSafe, when it is used for Web projects, a URL can be set as the deployment Web. This is where a Web project will be copied when it is finished and ready for production use. The Deploy command in Visual SourceSafe will cause the Project to be moved to the directory specified as the Deployment Web.

Directory Browsing When a request is sent to a Web server and there is no default page, the client will be presented a list of the contents of the folder specified by the URL (if directory browsing is allowed by the Web server). The client will be able to move up and down directory in the Web site.

DLL Dynamic Link Library. This is an executable program that is called by a program at runtime. Multiple users may be using the same copy of a DLL. It does not have to be compiled and linked with the calling program before usage.

DSN Data Source Name. The logical name given to a database connection in ODBC (Open Database Connectivity). The user of the database can refer to the DSN without having to refer to the actual database.

Encryption The technique of encoding data so that it is unusable to anyone lacking the ability to decrypt the data. This is a very important issue for sending sensitive data over the Internet, a very public medium.

FileView Tab In Visual InterDev, when a Workspace is open, the FileView Tab will be displayed at the bottom of the Workspace Window. When this tab is clicked, the projects and files in the projects will be displayed.

Flame When an individual has done something on a newsgroup or other public forum that upsets one or more users of the forum, he may be the target of messages that are uncomplimentary. This is called being "flamed." If it ever happens to you, it feels a little like kissing a blowtorch.

Forms Controls These are the ActiveX Controls that are the Controls available on the HTML Layout Control such as the Text box, Check box, and Command button.

FrontPage Microsoft's WYSIWYG editor and Web development tool. The FrontPage editor is included with Visual InterDev.

Full Text Search Searching all of the text in a document or HTML page for a word or phrase regardless of where in the text it appears.

GIF A file format for graphic images. GIF is supported by most Web browsers.

Global.ASA An HTML file that contains scripting instructions that are processed at the beginning or end of an application or session. There can be only one Global.ASA file per Web. It must be located on the root of the Web.

Home Page The starting page for a Web. Usually it is a default page and may be named Default.HTM, Default.ASP, or Index.HTM.

HTM The file extension for an HTML page.

HTML Layout Control An ActiveX Control that has its properties stored in an .ALX file. It is used to precisely control the location of command buttons, text boxes, and other form controls.

Hyperlinks A link or pointer to another file or document. When a hyperlink is clicked, the browser sends a request to the server at the URL for the file named in the hyperlink.

Identity In SQL Server an identity variable is the value assigned to a column that always contains a unique value. The value is assigned by the database engine when a new record is inserted into the table.

Image Composer Microsoft Image Composer is a graphic file editor. It supports several graphic file formats including .MIC which is the native format for Image Composer. It uses sprites to manipulate images.

Inbound Link A Hyperlink that points to the subject file or Web site.

Increment As used in SQL Server, it is the amount an identity variable is increased. As used in Java, JavaScript, JScript, and C++ it is the ++ operator which is used to increase a variable by 1.

InfoView Tab In Developer Studio (Visual InterDev) the Workspace window displays an InfoVeiw tab. This tab displays the documentation available in the IDE (Integrated Design Environment).

In-line error handling Error handling code is located at the point in the code where the error occurs rather than being in an error handling subroutine.

Insert Query A SQL statement that is used to insert a new record into a relational database table.

Internet Information Server The Web Server from Microsoft that is used on the Windows NT 4 Server platform. It is the industrial-strength Web server from Microsoft.

Intrinsic HTML Objects The objects that are created using only HTML code (text box or command button for example) are Intrinsic HTML objects.

Java A programming language based on C++. The Java program is compiled into bytecode, which is interpreted by another program called a Java Virtual Machine.

Java Virtual Machine A program that is specific to a particular system/OS platform. It interprets Java bytecode programs. This gives Java programs platform independence.

JavaScript A scripting language used in HTML Web pages that is loosely based on the Java language.

JPEG A graphics file format. It is supported by most Web browsers.

JPG The file extension for a graphic image format that is used extensively on Web pages on the Internet. It has the advantages of being supported by most Web browsers and being a compressed format.

JScript Microsoft's implementation of JavaScript.

LAN Local Area Network. A LAN is a direct connection between systems by cable (or radio for wireless LANs). The connection is not switched as in a connection over phone lines.

Links Also called hyperlinks. A link or pointer to another file or document. When a link is clicked, the browser sends a request to the server at the URL for the file named in the link.

LinkView A graphical display of the links between documents and files in a Web site. LinkView also tests the links to determine whether the link valid.

Loosely Typed Language In a loosely typed language, variables do not have to be declared with a data type before they are used. The data type will be assigned based on the data used to initialize the variable.

MDI Application Multiple Document Interface application. An MDI application is composed of a parent window or document which serves as a container for one or more child windows. Microsoft Word is an example of an MDI application where several document windows may be open simultaneously.

Media Manager Microsoft Media Manager is a content management system. It simplifies the storage and location of media files. Media Manager is integrated with the Windows Explorer. Media Manager provides a Thumbnail view.

Memory Leak A memory leak occurs when a memory allocation by a running program is not released when the program is closed.

MIC The file extension for the native image format of graphic files in Microsoft Image Composer. The image is composed of a background and one or more sprites. See *Sprites*.

Microsoft Access A desktop relational database management system that includes its own User Interface and report generation capabilities.

MID The file extension for a MIDI sound file. It is supported by most Web browsers. Microsoft Music Producer can save files in this format for use in Web pages.

MIDI Musical Instrument Digital Interface. A music industry protocol that defines the exchange of musical information between computers and instruments.

MMP The file extension for the native sound file format for Microsoft Music Producer.

Microsoft Image Composer Format The native format for Microsoft Image Composer. In this format, images are separate objects from the background. The image can be manipulated separately and is called a sprite.

Multi-Homed Servers In Microsoft Internet Information Server, a multi-homed server has multiple IP addresses assigned to the Network Interface Card (NIC) and multiple domain names that are served.

Multi-Tier Applications A multi-tier application is composed of layers or tiers. An example is an application with a user interface layer running on a group of systems. The UI could be Web browsers. The next layer is one or more systems in a Web server layer that are processing the interface to the UI layer and communication with a third layer that is composed of database servers.

Music Producer A Microsoft application that is used to create royalty free original musical compositions for use as background music for Web pages.

Netiquette The rules of good manners on the Internet. Violations of Netiquette are usually responded to with a flame (See *flame*). An example of a violation of Netiquette is called spamming which consists of sending e-mail to masses of recipients regardless of their desire to receive the e-mail. It is the Internet equivalent of junk mail.

Object An object is an entity that is created at runtime based on a class definition. Most objects have properties and methods.

Object-Oriented Language A programming language that is able to create class definitions and therefore objects.

OCX OLE Control. A windows control such as a text box that is based on the OLE model.

ODBC Open Database Connectivity. A set of middleware that provides an interface between User Interface programs and relational databases. The UI sends SQL statements to ODBC which submits the statement to the database engine. The database engine returns a results set to ODBC which passes it to the UI program.

ODBC Data Source The logical name associated with the connection to a database in the ODBC administrator.

OLE Object Linking and Embedding. An object in one process (the client object) can communicate with the server object in another process.

Outbound Link A hyperlink that points from the subject file to another file or Web site.

Peer Web Services The Windows NT 4 Workstation Web server. It is virtually identical to the Microsoft Internet Information Server for Windows NT 4 Server.

PERL Practical Extraction and Reporting Language. PERL is a scripting language that is used extensively for server-side programming in the Common Gateway Interface.

Personal Web Server The Microsoft Web server for Windows 95.

Project In Visual InterDev, it is the files and data connections that comprise a Web site.

Protocol Stack The programs and functions that provide the communication between an application and a transport medium such as an Ethernet cable.

Proxy Server A server that receives Internet requests from a system on a LAN and passes the request to an Internet connection, waits for the response, and passes the response back to the client system. This proxy arrangement creates a firewall between the LAN and the Internet.

RDBMS Relational Database Management System. The set of programs that manage the data in a relational database.

Relational Database Data organized in tables that are composed of rows and columns.

Repetitive Processing Repeating the same set of program instructions for a given number of times or until a condition changes. Also called a loop.

Sample Application Wizard The Visual InterDev Wizard that is used to install the sample applications supplied with Visual InterDev.

Secure Sockets Layer A function that manages the encryption and decryption of data to secure sensitive data on the very public medium of the Internet.

Seed The beginning number for an Identity variable in SQL Server.

Select Query A SQL statement used to retrieve data from a relational database table.

Shadow Folder In Visual SourceSafe, a Web project is in a Web folder that is used for test and development. This is called the Shadow Folder. When the project is ready for production it is deployed to the deployment folder.

Sprite In Microsoft Image Composer, an image file is composed of a sprite, which is the image and a background.

SQL Structured Query Language. The standardized language used to manipulate the data in a relational database.

SQL Server Database Device The physical file that contains one or more SQL Server database.

Subtypes (Data) In VBScript the only data type is Variant. Variant has data subtypes which are very much like the Visual Basic data types such as string and integer.

System Data Source An Open Database Connectivity (ODBC) data source that is available to all users on a system.

TCP Stack Transmission Control Protocol Stack. The programs that communicate between an application and a network transport medium such as a LAN cable.

Thumbnail Image A small version of a graphics file that can be viewed in Windows Explorer when Microsoft Media Manager is running.

To Do List (FrontPage) A mini project management tool for use by members of a development team using FrontPage. A list of tasks can be kept and is visible to all team members.

Transaction Processing In SQL Server, a change to records in a relational database table may involve several SQL statements and tables. If all of the changes are not completed successfully, none of the changes should be made. This set of SQL statements is grouped into a transaction. If all complete successfully the transaction is committed, if not all changes are rolled back.

UI Component User Interface Component. An application program that provides the interface to the user for an application. The functions usually involved are the display of data and the entry of data.

Update Query A SQL statement that alters the contents of an existing relational database record.

URL Universal Resource Locator. The address of a file or Web site. http://www.microsoft.com is an example.

Variable Coercion The conversion of a JavaScript of JScript variable's data type by an action such as including a number in a string variable.

Variant The data type for all variables in VBScript is Variant. The Variant data type has sub types such as string and integer.

VBScript A scripting language that is based on Visual Basic.

Virtual Directory A directory that is logically included in a Web site regardless of the physical location.

Virtual Server A server can have more than one IP address and domain name assigned to it. The additional IP addresses are virtual servers. The server is said to be multi-homed.

Visual Basic 5.0 Control Creation Edition The version of Visual Basic that is used for the creation of ActiveX Controls.

Visual C++ A full-featured compiled, programming language that uses class definitions to create runtime objects.

Visual SourceSafe A source code version control and library system from Microsoft that is integrated with Visual InterDev.

W3C World Wide Web Consortium. A group that has been assigned the task of creating standards for the World Wide Web.

WAN Wide Area Network. This will usually involve switched telephone connections or leased lines. The range of a WAN can be world wide.

WebBot A WebBot component is a dynamic object on a page that is evaluated and executed when the page is saved or when a browser loads the page. Most WebBot components generate HTML.

Workspace Visual InterDev keeps the set of projects that are being worked on in a Workspace.

Index

Symbols

& concatenation operator (VBScript), 84
401(k) sample application (companion CD), 62

A

Access (Microsoft), 288
Active Server, 66
  functions, TCP/IP Stack, 67
  Global.asa file, 52
  installing, 67, 278-279
  JScript, HTML forms, 99-102
  multi-tier applications, 118
  Web page transmission, 68
Active Server pages, 26, 283-284
  adding to sample project, 251-252
    HTML code, 252-254
  creating, 68-73, 164-168
    program logic, adding, 72
    VBScript source code, 73
  databases, connecting to, 257-259
  HTML forms
    gathering user information, 86-90
    source code, 89-90
  testing, 73-75
ActiveX, 284
  ADO (ActiveX Data Object), 163-168
  ClassID, 285
  COM, 285
  defining, 105-106
  Forms Control, 286
  HTML code, 261-263
  HTML Layout Control, 287
ActiveX Controls, 106
  HTML Layout Wizard, 219-220
  inserting into Web Pages, 27, 106-110
    testing HTML pages, 109-110
  purchasing (Web sites), 108
  Visual Basic 5.0 Control Creation Edition, 111-113
ADC (Advanced Data Connector), 168
addresses (Internet), 6
  IP (Internet Protocol), 7
  URL, 291
ADO (ActiveX Data Object), 163-168
Advanced Data Connector (ADC), 168
Advanced Research Planning Agency Net (ARPANET), 3-4
ALX file, 284
anonymous access to Web sites, 204
applets (Java), 93
  Java Virtual Machine, 14
applications
  MDI (Multiple Document Interface), 288
  multi-tier, 115, 289
    Active Server, 118
    client/component communications, 116-117
    DCOM (Distributed Component Object Model), 116-118
  sample applications (companion CD)
    401(k), 62
    DosPerros Tutorial Application, 59-61
arguments, 211
ARPANET (Advanced Research Planning Agency Net), 3-4
Array objects (JScript), 98
arrays (VBScript), 81
ASA (file extension), 284
ASCII text, 284
ASP (file extension), 284
assignment operators (JScript), 96

B

bitwise operators (JScript), 96
Boolean (JScript data types), 96
bottleneck functions, 284
broken links, 284
browsers, 12-14
  security, 261
browsing directory, 286
Button control (HTML), 210-211
bytecode, 284

C

C++, 284
Call Stack window (Script Debugger), 233
CD-ROM (included with book), 46, 281-282
  contents, 281-282
  legal agreement, 282
CERN (European Laboratory for Particle Physics), 18-19
CGI, 284
Change Password command (Visual SourceSafe Users menu), 192
Check Box control (HTML Layout Wizard), 214-216, 222
Check Hyperlinks command (Visual SourceSafe Web menu), 197
Check In command (SourceSafe menu), 197
Check Out command (SourceSafe menu), 196
Checked property (Check Box control), 214
Choose Name dialog box, 154
ClassID (ActiveX), 285
client scripting
  advantages, 216
  HTML forms
    Button control, 210-211
    Check Box control, 214-216
    Password control, 213
    Radio Button control, 213-214
    Reset control, 211
    Submit control, 212
    Text control, 212
client/component communications, 116-117
client/server computing, 18
client/server technology, 9-10
CLSID (ActiveX), 285
COBOL (Common Business Oriented Language), 17-18
code
  comments (VBScript), 86
  error handling, 287
Code Context, 285
Code window (Script Debugger), 232-233
coding conventions (VBScript), 85-86

COM (ActiveX), 285
Combo box control (HTML Layout Wizard), 222
Command button control (HTML Layout Wizard), 222
commands
  File menu
    New, 26, 48, 106
    Open SourceSafe Database (Visual SourceSafe), 193
    Open Workspace, 55
    Save As, 143
    Set Working Folder (Visual SourceSafe), 194
  Insert menu, New Database Item, 140, 153
  SourceSafe menu
    Check In, 197
    Check Out, 196
  Start menu, Internet Service Manager, 268
  Tools menu
    Customize, 36, 39
    Show Differences (Visual SourceSafe), 200
    Show History (Visual SourceSafe), 199
    Show To Do List, 187
    View Links, 178
  Users menu, Change Password (Visual SourceSafe), 192
  View menu
    HTML, 183
    Output, 32
    Source, 216
  Web menu
    Check Hyperlinks (Visual SourceSafe), 197
    Create Site Map, 198
    Deploy (Visual SourceSafe), 195
commenting code (VBScript), 86
Common Business Oriented Language (COBOL), 17-18
communications (client/component), 116-117
companion CD, 46
  contents, 281-282
  legal agreement, 282
compiling, 285
composing music (Music Producer), 243
concatenation operators (VBScript), 84
conditional logic, 285
conditional statements (JScript), 96-97
constants (VBScript), 81-82
controls (HTML)
  Button, 210-211
  Check Box, 214-216
  Password, 213
  Radio Button, 213-214
  Reset, 211
  Submit, 212
  Text, 211-212
copyrights, 237
Create New Data Source dialog box, 137, 156
Create Site Map command (Visual SourceSafe Web menu), 198
Customize command (Tools menu), 36, 39
Customize dialog box, 36
customizing
  HTML Layout Wizard, 221
  toolbar, 37-38
  Tools menu, 40-43

D

data, encrypting, 286
Data Form Wizard, 25
Data Source Name (DSN), 161-164
data subtypes, 290
data types
  JScript, 95-96
  VBScript, 79-80
database devices, creating, 151-152
database engines, 285
Database Project Wizard, 25
database tools, 27-28
databases
  creating, 149-155
    database devices, 151-152
    SQL Server logging, 152
    tables, 153-155
  DSN (Data Source Name), 286
  normalization, 125-126
    constructing normalized databases, 126-129
  Open Database Connectivity (ODBC), 136-139
  projects, creating for existing databases, 155-158
  queries, 135-136
    delete queries, 146-147
    insert queries, 145-146
    select queries, 139-143
    update queries, 144
  relational, 290
    SQL (Structured Query Language), 129-133, 290
    tables, 124-125
DataView tab, 285
DCOM (Distributed Component Object Model), 116-117, 285
  application scaling, 117-118
DEC VT-100, 285
declaring
  constants (VBScript), 82
  variables (JScript), 95
default pages, 285
Default.HTM files, 224, 249-250
  examining, 250-251
delete queries, designing, 146-147
Delete Query, 136, 285
DELETE statements (Structured Query Language), 132
Departmental Site Wizard, 25
Deploy command (Visual SourceSafe Web menu), 195
deploying Web projects (Visual SourceSafe), 195-196
Deployment Web, 286
dialog boxes
  Choose Name, 154
  Create New Data Source, 137, 156
  Customize, 36
  Insert ActiveX Control, 107
  New, 136
  New Device Information, 152
  ODBC SQL Server Setup, 157
  ODBC SQL Setup, 138
  Properties (Personal Web Server), 277
  Query Designer, 140-141
  Select Data Source, 136, 156
directories
  root directories, 205-206
  virtual, 291
  Web virtual directories, 71
directory browsing, 286
Distributed Component Object Model (DCOM), 116-117
  application scaling, 117-118
DLL (Dynamic Link Library), 286
DNS (Domain Name Servers), 7
docking toolbars, 35-36
documents, adding to projects, 26-27
Domain Name Servers (DNS), 7
domain names, 6

DosPerros Tutorial Application (companion CD), 59-61
downloading Internet Explorer, 5
DSN (Data Source Name), 161-164, 286

E

e-mail, 4
Edit window (Script Debugger), 231
editing project files, 52, 57
Editor (FrontPage 97), 182-183, 286
encryption, 286
  Secure Sockets Layer, 49, 290
engines, database, 285
error handling, in-line, 287
European Laboratory for Particle Physics (CERN), 18-19
eval() function (JScript), 97
events, OnClick
  Radio Button control, 214
  Reset control, 211
Excel charts, 26
Excel worksheets, 27
Execute access, Web site security, 207
Explorer, *see* Internet Explorer

F

file extensions
  ALX, 283
  ASA, 283-284
  ASP, 283-284
  GIF, 283
  HTM, 283, 286
  JPG, 283, 287
  MIC, 283, 288
  MID, 283, 288
  MMP, 283, 288
file formats
  GIF, 286
  images, 238-239
    selecting, 239
  JPEG, 287
  music, 243-244
File menu commands
  New, 26, 48, 106
  Open Workspace, 55
  Save As, 143
  Visual SourceSafe
    Open SourceSafe Database, 193
    Set Working Folder, 194
File Transfer Protocol (FTP), 4
files
  Active Server Pages, 26
  adding to projects, 26, 55-56
  Global.asa, 52, 71
  Help, 32-34
  HTML layout, 26
  HTML pages, 26
  macros, 26
  multimedia (Visual InterDev tools), 28
  ODBC scripts, 26
  organizing with Media Manager, 244-246
  projects
    editing, 52, 57
    Web server folders, 53-54
    working copies, 52-53
  text, 26
  viewing, 50
  Working Folder (Visual SourceSafe), 194
FileView tab, 286
firewalls, proxy servers, 290
first normal form (normalization), 125
flaming, 286
floating toolbars, 35-36
folders
  MyProjects, 52
  Web server (project files), 53-54
foreign keys, 128
forms (HTML)
  JScript, 99-102
  VBScript, 86-90
Forms Control (ActiveX), 286
four event handling subroutines (VBScript), 256
FrontPage, 27, 286
  Editor, 182-183
  Server Extensions, installing, 279
  To Do List, 290
  Web pages, sharing with Visual InterDev, 184-187
  WebBots, 182
FTP (File Transfer Protocol), 4
full text searches, 50, 286
Function procedure (VBScript), 84-85
functions
  JScript, 97
  VBScript, 80

G

GIF, 286
Global.asa files, 52, 71, 283-284
Gopher, 4
graphic browsers, 13
graphic file formats
  GIF, 286
  JPEG, 287

H

Help functions, 32-34
Help tutorial (Image Composer), 242
hiding toolbars, 36-37
home pages, 286
Hot spot control (HTML Layout Wizard), 223
HTM (file extension), 283, 286
HTML (HyperText Markup Language)
  defined, 19
  history, 18-19
  limitations, 78-79
  standards, 20-21
  versions, 20-21
HTML code
  Active Server Page, 252-254
  ActiveX, 261-263
  Default.HTM, 250-251
  Search.HTM page, 263
HTML command (View menu), 183
HTML documents
  Link View, 174-176
  links, 174
HTML forms
  controls
    Button, 210-211
    Check Box, 214-216
    Password, 213
    Radio Button, 213-214
    Reset, 211
    Submit, 212
    Text, 211-212
  JScript, 99-102
    collecting user information, 99-102
    source code, examining, 100-102
  VBScript, 86-90
    collecting user information, 86-89
    source code, examining, 89-90
HTML Insertion Wizard, 25
HTML Layout Control (ActiveX), 283-284, 287
  code, 259-260
  scripting, 260
HTML layout files, 26
HTML Layout Wizard, 219-220
  creating HTML layouts, 226-227
  customizing, 221

Default.HTM file, 224-225
Script Wizard, 224
Template Page Wizard, 220-223
testing layouts, 225-226
HTML objects, intrinsic, 287
HTML page files, 26
HTML tags, 20
hyperlinks, 287
checking with Visual SourceSafe, 197

I

IDC (Internet Data Connector), 168
IDE (Integrated Design Environment), 287
identity variables, 287
increments, 287
seed, 290
Image Composer, 28, 287
adding to Tools menu, 40-43
features, 238
format, 288
Help tutorial, 242
image file formats, 238-239
selecting, 239
installing, 280
Sprites, 239-241, 290
Tools menu, 241
Arrange palette, 242
Image control (HTML Layout Wizard), 223
Immediate window (Script Debugger), 233
In Context Help function, 34
in-line error handling, 287
inbound links, 287
increments, 287
InfoView tab, 287
initiating Script Debugger, 231
Insert ActiveX Control dialog box, 107
Insert menu commands, New Database Item, 140, 153
insert queries, designing, 145-146
Insert Query, 136
SQL, 287
INSERT statements (Structured Query Language), 133
installing
401(k) sample application (companion CD), 62
Active Server, 67, 278-279
DosPerros Tutorial Application, 59-61
Image Composer, 280
Media Manager, 280
Music Generator, 280
Script Debugger, 230-231
Server Extensions (FrontPage), 279
SQL Server Service Pack 2, 150
Visual InterDev Client, 279-280
Visual SourceSafe, 190-192
Windows 95 Personal Web Server, 276-277
Properties Dialog Box, 277
Windows NT 4 Internet Information Server, 278
Internet
addresses, 6
IP (Internet Protocol), 7
URL, 291
client/server technology, 9-10
development, 10-14
browsers, 12-14
platforms, 11
servers, 11-12
domain names, 6
e-mail, 4
FTP (File Transfer Protocol), 4
Gopher, 4
history, 3-4
Netiquette, 289
news groups, 5
services, 4-6
spamming, 289
Telnet, 5
World Wide Web, *see* WWW
Internet Data Connector (IDC), 168
Internet Explorer, 13
downloading, 5
Script Debugger
Call Stack window, 233
Code window, 232-233
Edit window, 231
Immediate window, 233
initiating, 231
installing and uninstalling, 230-231
Project Explorer window, 232
Web site, 229
VBScript, 21
Internet Information Server, 287
installing (Windows NT 4), 278
Internet Protocol (IP), 7
InterNIC Web site, 7
intranets, 9-10
developing Web pages
browsers, 12-14
platforms, 11
servers, 11-12
development, 10-15
intrinsic controls (HTML), 211
intrinsic objects
HTML, 287
JScript, 97-98
IP (Internet Protocol), 7
ISAPI support (MS IIS), 267
IUSR_ServerName account, 204

J-K

Java, 91-93, 287
applets, 93
bytecode, 284
defining, 92-93
Java Virtual Machine, 14, 287
JavaScript, 21, 287
variables, coercion, 291
joining tables (Structured Query Language), 130-131
JPEG, 287
JPG (file extension), 287
JScript, 94-99, 288
Active Server, HTML forms, 99-102
data types, 95-96
functions, 97
objects
creating, 98-99
intrinsic, 97-98
Operator Precedence, 96
operators, 96
program flow statements, 96-97
variables, 95
coercion, 291
writing code, 94

L

Label control (HTML Layout Wizard), 222
LAN (Local Area Network), 288
proxy servers, 290
languages
C++, 284
compiling, 285
Java, 287
bytecode, 284
JavaScript, 287
loosely typed, 288
object-oriented, 289

PERL (Practical Extraction and Reporting Language), 289
SQL (Structured Query Language), 290
VBScript, 291
Visual C++, 291
Link View, 174-176
toolbar, 177-178
viewing Web sites, 178
links, 174
broken, 284
checking, 57-58
at Web site, 263
hyperlinks, 287-288
inbound, 287
outbound, 289
LinkView, 27, 288
List box control (HTML Layout Wizard), 222
logging (SQL Servers), 152
logic, conditional, 285
logical operators (JScript), 96
loops, 290
loosely typed languages, 288
LYNX browser, 12-13

M

macro files, 26
MDI (Multiple Document Interface) applications, 288
Media Manager (Microsoft), 28, 288
features, 244-246
installing, 280
locating, 246
memory leaks, 288
MIC, 288
Microsoft
Access, 288
Active Server, 66
functions, TCP/IP Stack, 67
Global.asa file, 52
installing, 67, 278-279
JScript, HTML forms, 99-102
multi-tier applications, 118
Web page transmission, 68
see also Active Server pages
FrontPage, 286
Image Composer, 287
features, 238
format, 288
Help tutorial, 242
image file formats, 238-239
Sprites, 239-241, 290
Tools menu, 241-242
Internet Explorer, 13
downloading, 5
Script Debugger, 229-233
VBScript, 21
Internet Information Server, 287
JScript, *see* JScript
Media Manager, *see* Media Manager
Music Generator, 28
installing, 280
Music Producer, 283, 289
composing, 243
features, 242
file formats, 243-244
sound, changing, 243
Visual InterDev Client, installing, 279-280
Visual SourceSafe, *see* Visual SourceSafe
Web servers, 12
Web site
ActiveX Controls, 108
Internet Explorer, downloading, 5
MS IIS 3.01, downloading, 67
technical support, 275
Verisign registration and certification, 110
Web tools, 23-24
MID (file extension), 283, 288
MIDI (Musical Instrument Digital Interface), 288
sound files, 283
MILNET, 4
MMP (file extension), 283, 288
Mosaic, 13
MS IIS (Microsoft Internet Information Server)
anonymous access to Web sites, 204
directories, 270
editing, 271-272
Internet Service Manager, 268-269
ISAPI support, 267
limiting access, 273
logging, 272
WWW Service Properties dialog box, 269
multi-homed servers, 288
multi-tier applications, 289
Active Server, 118
client/component communications, 116-117
DCOM (Distributed Component Object Model), 116-118
defined, 115
multi-tier ODBC drivers, 164
multimedia files (Visual InterDev tools), 28
Music Generator, 28
installing, 280
Music Producer, 283, 289
composing, 243
features, 242
file formats, 243-244
sound, changing, 243
MyProjects folder, 52

N

NAME argument (Button control), 211
Name property (Text control), 212
naming conventions (VBScript), 85-86
netiquette, 5, 289
Netscape Navigator, 13
JavaScript, 21
Netscape Web servers, 11
New command (File menu), 26, 48, 106
New Database Item command (Insert menu), 140, 153
New Database Wizard, 25, 149-153
database devices, creating, 151-152
New Device Information dialog box, 152
New dialog box, 136
news groups, 5
news readers, downloading, 5
normal form, 125
normalization, 125-126
constructing normalized databases, 126-129
numbers (JScript data types), 96

O

Object Linking and Embedding (OLE), ActiveX, 105-106
object-oriented languages, 289
objects, 289
creating in JScript, 98-99
intrinsic
HTML, 287
JScript, 97-98
WebBot, 291
OCX, 289

ODBC (Open Database Connectivity), 160, 289
  Active Server pages, creating, 164-168
  ADO, 164-168
  creating database projects, 155-158
  database connections, 136-139, 155-158
  drivers, 161, 164
  DSN, 161-164
  System Data Source, 290
ODBC script files, 26
ODBC SQL Server Setup dialog box, 157
ODBC SQL Setup dialog box, 138
OLE (Object Linking and Embedding), 289
  ActiveX, 105-106
OnClick event
  Radio Button control, 214
  Reset control, 211
Open SourceSafe Database command (Visual SourceSafe File menu), 193
Open Workspace command (File menu), 55
opening workspaces, 47
Operator Precedence
  JScript, 96
  Visual Basic, 84
operators
  JScript, 96
  VBScript, 84
Option button control (HTML Layout Wizard), 222
outbound links, 289
Output command (View menu), 32
Output window, 32

P

packets, 3
Password control (HTML), 213
Peer Web Services, 289
PERL (Practical Extraction and Reporting Language), 289
Personal Web Server, 289
  directories, 270
    editing, 271-272
  installing (Windows 95), 276-277
  logging, 272
  managing, 268-269
  Properties Dialog Box, 277
platforms, 11
PowerPoint files, 27
Preview tool, 27
procedures (VBScript), 84-85
program flow control statements
  JScript, 96-97
  VBScript, 82-83
program logic, adding to Active Server Pages, 72
Project Explorer window (Script Debugger), 232
projects, 46, 289
  Active Server Page
    adding, 251-252
    HTML code, 252-254
  ActiveX, HTML code, 261-263
  creating, 48-54, 248-249
    Web structures, 49-52
  databases
    connecting to, 257-259
    creating for existing databases, 155-158
  Default.HTM page
    adding, 249-250
    examining, 250-251
  files
    adding, 55-56
    editing, 52, 57
    viewing, 50
    Web server folders, 53-54
    working copies, 52-53
  HTML Layout Control
    code, 259-260
    scripting, 260
  links, checking, 263
  removing, 58
  Search.HTM page, HTML code, 263
  specifications, 248
properties
  Checked (Check Box control), 214
  Name (Text control), 212
  Value (Text control), 212
Properties dialog box (Personal Web Server), 277
Properties tool, 27
protocol stacks, 116, 289
protocols
  Gopher, 4
  TCP/IP (Transfer Control Protocol/Internet Protocol), 6-7
proxy servers, 290

Q

queries, 135-136
  deleting, 146-147
  inserting, 145-146
  selecting, 139-143
  updating, 144
Query Designer, 27
Query Designer dialog box, 140-141

R

Radio Button control (HTML), 213-214
RDBMS (Relational Database Management System), 290
  ODBC (Open Database Connectivity), 160-161
relational databases, 290
  normalization, 125-126
    constructing normalized databases, 126-129
  Structured Query Language (SQL), 129-133, 290
    DELETE statement, 132
    INSERT statement, 133
    joining tables, 130-131
    SELECT statement, 129-130
    UPDATE statement, 131-132
  views, 131
  tables, 124-125
removing Web projects, 58
repetitive processing, 290
Reset control (HTML), 211
Results List window, 33
root directories, 205-206
royalties, 237

S

Sample Application Wizard, 25, 59-62, 290
Save As command (File menu), 143
Script Debugger
  Call Stack window, 233
  Code window, 232-233
  Edit window, 231
  Immediate window, 233
  initiating, 231
  installing and uninstalling, 230-231
  Project Explorer window, 232
  Web site, 229
Script Wizard, 25, 224, 256-257
scripting languages, 21, 78-79
Scroll bar control (HTML Layout Wizard), 223
Search.HTM page, HTML code, 263

searching (full text searches), 50
second normal form (normalization), 126
Secure Sockets Layer, 49, 290
security
  anonymous access, 204
  browsers, 261
  Execute access, 207
  MS IIS, 204
  root directories, 205-206
  source code control, 190
    Visual SourceSafe, 192-193
  Windows 95, 205
  Windows NT, 204-205
seed, 290
Select Case statement (VBScript), 83
Select Data Source dialog box, 136, 156
select queries, designing, 139-143
  Query Designer dialog box, 140-141
Select Query, 136, 290
SELECT statement (Structured Query Language), 129-130
Server Extensions (FrontPage), installing, 279
servers, 11-12
  identifying by name, 70
  Internet Information Server, 287
  Microsoft Web servers, 12
  MS IIS
    directories, 270-272
    Internet Service Manager, 268-269
    limiting access, 273
    logging, 272
    WWW Service Properties dialog box, 269
  multi-homed, 288
  Netscape Web servers, 11
  Peer Web Services, 289
  Personal Web Server, 289
    directories, 270
    directories, editing, 271-272
    logging, 272
    managing, 268-269
  proxy, 290
  virtual, 291
  Windows NT Security, 269
Set Working Folder command (Visual SourceSafe File menu), 194
SGML (Standard Generalized Markup Language), 19
Shadow Folder, 195, 290
Show Differences command (Visual SourceSafe Tools menu), 200
Show History command (Visual SourceSafe Tools menu), 199
Show To Do List command (Tools menu), 187
showing toolbars, 36-37
single-tier ODBC drivers, 164
site maps, creating in Visual SourceSafe, 198
sound, changing in Music Producer, 243
sound files, MID file extension, 288
source code control (Visual SourceSafe), 190
  checking out projects, 196
  deploying projects, 195-196
  hyperlinks, checking, 197
  installing, 190-192
  project history, viewing, 199
  project security, 192-193
  reconciling file copies, 196
  Shadow Folder, 195
  Show Differences tool, 200
  site maps, creating, 198
  Visual SourceSafe database, 193
  Web projects, 194
  Working Folder, 194
Source command (View menu), 216
SourceSafe menu commands
  Check In, 197
  Check Out, 196
spamming, 289
Spin control (HTML Layout Wizard), 223
Sprites, 239-241, 290
SQL (Structured Query Language), 129-133, 290
  joining tables, 130-131
  ODBC, 160
    Active Server pages, creating, 164-168
    ADO, 164-168
    drivers, 161, 164
    DSN, 161-164
  statements
    DELETE, 132
    Delete Query, 285
    INSERT, 133
    Insert Query, 287
    SELECT, 129-130
    Select Query, 290
    UPDATE, 131-132
    Update Query, 291
  views, 131
SQL Server
  Database Device, 290
  databases
    creating, 149-155
    database devices, creating, 151-152
    tables, creating, 153-155
  identity variables, 287
    increments, 287
  logging, 152
  ODBC database connections, 136-139
  seed, 290
  transaction processing, 291
SQL Server Service Pack 2, installing, 150
Standard Generalized Markup Language (SGML), 19
Start menu commands, Internet Service Manager, 268
String objects, 98
strings (JScript data types), 95
Structured Query Language, *see* SQL
Sub procedure (VBScript), 84-85
Submit control (HTML), 212
subtypes (data), 290
System Data Source, 290

T

Tab strip control (HTML Layout Wizard), 223
tables, 124-125
  creating in SQL Server databases, 153-155
  foreign keys, 128
  joining (Structured Query Language), 130-131
  normalization, 125-126
    constructing normalized databases, 126-129
tags (HTML), 20
TCP (Transfer Control Protocol), 6-7
  TCP Stack, 290
TCP/IP, 6-7
  TCP/IP Stack, 67
Telnet, 5
Template Page Wizard (HTML Layout Wizard), 220-223
Template Wizard, 25
terminal emulations, DEC VT-100, 285
testing
  Active Server pages, 73-75
    ASP version, 73
    Browser version, 74
    HTML version, 75
  HTML layouts, 225

text, full text searches, 50, 286
Text box control (HTML Layout Wizard), 222
Text control (HTML), 211-212
text files, 26
  ALX file, 284
text only browsers, 12-13
third normal form (normalization), 126
thumbnail images, 290
To Do List (FrontPage), 187, 290
Toggle button (HTML Layout Wizard), 222
toolbars, 34-40
  adding tools, 37-38
  creating, 39-40
  docking, 35-36
  floating, 35-36
  hiding/showing, 36-37
  Link View, 177-178
tools (Visual InterDev), 24-28
  database tools, 27-28
  document creation, 26-27
  file creation, 26-27
  multimedia file management, 28
  Wizards, 25
Tools menu
  adding tools, 40-43
  Image Composer palettes, 241-242
Tools menu commands
  Customize, 36, 39
  Show To Do List, 187
  View Links, 178
  Visual SourceSafe
    Show Differences, 200
    Show History, 199
transaction processing, 291
Transfer Control Protocol (TCP), 6-7
  TCP Stack, 290
Transfer Control Protocol/Internet Protocol (TCP/IP), 6-7
  TCP/IP Stack, 67
TYPE argument (Button control), 211

U

UI (User Interface) Component, 291
update queries, designing, 144
Update Query, 136, 291
UPDATE statements (Structured Query Language), 131-132
URL (Universal Resource Locator), 291
Use Groups, 5
Users menu commands (Visual SourceSafe), Change Password, 192

V

VALUE argument (Button control), 211
Value property (Text control), 212
variables
  Code Context, 285
  coercion, 291
  identity, 287
    increments, 287
    seed, 290
  JScript, 95
  VBScript, 80-81
Variant data type (VBScript), 79-80, 291
VBScript, 21, 77-78, 291
  Active Server Pages, 73
  arrays, 81
  coding conventions, 85-86
  commenting code, 86
  concatenation operators, 84
  constants, 81-82
  data subtypes, 290
  data types, 79-80, 291
  four event handling subroutines, 256
  functions, 80
  HTML controls
    Button, 210-211
    Check Box, 214-216
    Password, 213
    Radio Button, 213-214
    Reset, 211
    Submit, 212
    Text, 211-212
  HTML forms, 86-90
    source code, 89-90
  naming conventions, 85-86
  operators, 84
  procedures, 84-85
  program flow control statements, 82-83
  variables, 80-81
VBScript code,connecting to databases, 257-259
View Links command (Tools menu), 178
View menu commands
  HTML, 183
  Output, 32
  Source, 216
viewing
  files, 50
  Web project history (Visual SourceSafe), 199
  Web sites, 178
views (Structured Query Language), 131
virtual directories, 291
virtual servers, 291
Visual Basic, 291
  Operator Precedence, 84
Visual Basic 5.0 Control Creation Edition, 111-113
Visual C++, 291
Visual InterDev Client, installing, 279-280
Visual InterDev tools, 24-28
  database tools, 27-28
  document creation, 26-27
  file creation, 26-27
  multimedia file management, 28
  Wizards, 25
Visual J++, 91-92
Visual SourceSafe, 291
  checking out projects, 196
  deploying projects, 195-196
  Deployment Web, 286
  hyperlinks, checking, 197
  installing, 190-192
  project history, viewing, 199
  project security, 192-193
  reconciling file copies, 196
  Shadow Folder, 195, 290
  Show Differences tool, 200
  site maps, creating, 198
  Visual SourceSafe database, 193
  Web projects, 194
  Working Folder, 194

W-Z

W3C (World Wide Web Consortium), 291
W3C Web site, 21
WAN (Wide Area Network), 291
Web addresses, 6
  IP (Internet Protocol), 7
Web browsers, 12-14
  graphic, 13
  Internet Explorer, 13
  Netscape Navigator, 13
  text only, 12-13
Web editors, 22

Web menu commands (Visual SourceSafe)
Check Hyperlinks, 197
Create Site Map, 198
Deploy, 195
Web pages
development, 10-14
browsers, 12-14
platforms, 11
servers, 11-12
inserting ActiveX Controls, 106-110
testing HTML pages, 109-110
links, checking, 57-58
sharing between FrontPage 97 and Visual InterDev, 184-187
see also Web sites
Web Project Wizard, 25, 48-49, 220, 248-249
Active Server Pages, creating, 68-71
creating projects, 48-54
Web structures, 49-52
Web projects, *see* projects
Web server folders (project files), 53-54
Web servers, 11-12
Web sites
anonymous access, 204
copyrights and royalties, 237
home pages, 286
Internet Explorer Script Debugger, 229
InterNIC, 7
links, checking, 263
LinkView, 288
Microsoft
ActiveX Controls, 108
Internet Explorer, downloading, 5
MS IIS 3.01, downloading, 67
technical support, 275
Verisign registration and certification, 110
security
Execute access, 207
root directories, 205-206
viewing, 178
virtual directories, 291
W3 Consortium, 21
see also Web pages
Web structures
creating (projects), 49-52
full text searches, 50
Web virtual directories, 71
WebBots (FrontPage 97), 182, 291
windows
Output, 32
Results List, 33
Windows 95, security, 205
Windows 95 Personal Web Server
installing, 276-277
Properties Dialog Box, 277
Windows NT, security, 204-205
servers, 269
Windows NT 4 Internet Information Server, installing, 278
Wizards, 25
Departmental Site, 25
New Database, 149-153
database devices, creating, 151-152
Sample Application, 59-62, 290
Script, 25, 224, 256-257
Web Project, 48-49, 248-249
Active Server Pages, creating, 68-71
creating projects, 48-54
Word documents, 26
Working Folder (Visual SourceSafe), 194
Workspace, 291
Workspace window, 26
workspaces
defined, 46-47
opening, 47
WWW (World Wide Web), 5-6

Technical Support:

If you cannot get the CD/Disk to install properly, or you need assistance with a particular situation in the book, please feel free to check out the Knowledge Base on our Web site at **http://www.superlibrary.com/general/support**. We have answers to our most Frequently Asked Questions listed there. If you do not find your specific question answered, please contact Macmillan Technical Support at **(317) 581-3833**. We can also be reached by email at **support@mcp.com**.